The Thermodynamics of Transitional States

Edited by Paul F. Kisak

Contents

Chapter 1

State of matter

Not to be confused with Phase (matter).

In physics, a **state of matter** is one of the distinct forms that matter takes on. Four states of matter are observable in everyday life: solid, liquid, gas, and plasma. Many other states are known to exist only in extreme situations, such as Bose–Einstein condensates, neutron-degenerate matter and quark-gluon plasma, which occur in situations of extreme cold, extreme density and extremely high-energy color-charged matter respectively. Some other states are believed to be possible but remain theoretical for now. For a complete list of all exotic states of matter, see the list of states of matter.

Historically, the distinction is made based on qualitative differences in properties. Matter in the solid state maintains a fixed volume and shape, with component particles (atoms, molecules or ions) close together and fixed into place. Matter in the liquid state maintains a fixed volume, but has a variable shape that adapts to fit its container. Its particles are still close together but move freely. Matter in the gaseous state has both variable volume and shape, adapting both to fit its container. Its particles are neither close together nor fixed in place. Matter in the plasma state has variable volume and shape, but as well as neutral atoms, it contains a significant number of ions and electrons, both of which can move around freely. Plasma is the most common form of visible matter in the universe.[1]

The term phase is sometimes used as a synonym for state of matter, but a system can contain several immiscible phases of the same state of matter (see Phase (matter) for further discussion of the difference between the two terms).

1.1 The four fundamental states

1.1.1 Solid

Main article: Solid

In a solid the particles (ions, atoms or molecules) are closely packed together. The forces between particles are strong so that the particles cannot move freely but can only vibrate. As a result, a solid has a stable, definite shape, and a definite volume. Solids can only change their shape by force, as when broken or cut.

In crystalline solids, the particles (atoms, molecules, or ions) are packed in a regularly ordered, repeating pattern. There are various different crystal structures, and the same substance can have more than one structure (or solid phase). For example, iron has a body-centred cubic structure at temperatures below 912 °C, and a face-centred cubic structure between 912 and 1394 °C. Ice has fifteen known crystal structures, or fifteen solid phases, which exist at various temperatures and pressures.[2]

Glasses and other non-crystalline, amorphous solids without long-range order are not thermal equilibrium ground states; therefore they are described below as nonclassical states of matter.

The four fundamental states of matter. Clockwise from top left, they are solid, liquid, plasma, and gas, represented by an ice sculpture, a drop of water, electrical arcing from a tesla coil, and the air around clouds, respectively.

A crystalline solid: atomic resolution image of strontium titanate. Brighter atoms are Sr and darker ones are Ti.

Solids can be transformed into liquids by melting, and liquids can be transformed into solids by freezing. Solids can also change directly into gases through the process of sublimation, and gases can likewise change directly into solids through deposition.

1.1.2 Liquid

Structure of a classical monatomic liquid. Atoms have many nearest neighbors in contact, yet no long-range order is present.

Main article: Liquid

A liquid is a nearly incompressible fluid that conforms to the shape of its container but retains a (nearly) constant volume independent of pressure. The volume is definite if the temperature and pressure are constant. When a solid is heated above its melting point, it becomes liquid, given that the pressure is higher than the triple point of the substance. Intermolecular (or interatomic or interionic) forces are still important, but the molecules have enough energy to move relative to each other and the structure is mobile. This means that the shape of a liquid is not definite but is determined by its container. The volume is usually greater than that of the corresponding solid, the best known exception being water, H_2O. The highest temperature at which a given liquid can exist is its critical temperature.[3]

1.1.3 Gas

Main article: Gas

A gas is a compressible fluid. Not only will a gas conform to the shape of its container but it will also expand to fill the container.

In a gas, the molecules have enough kinetic energy so that the effect of intermolecular forces is small (or zero for an ideal gas), and the typical distance between neighboring molecules is much greater than the molecular size. A gas has no definite shape or volume, but occupies the entire container in which it is confined. A liquid may be converted to a gas by heating at constant pressure to the boiling point, or else by reducing the pressure at constant temperature.

At temperatures below its critical temperature, a gas is also called a vapor, and can be liquefied by compression alone

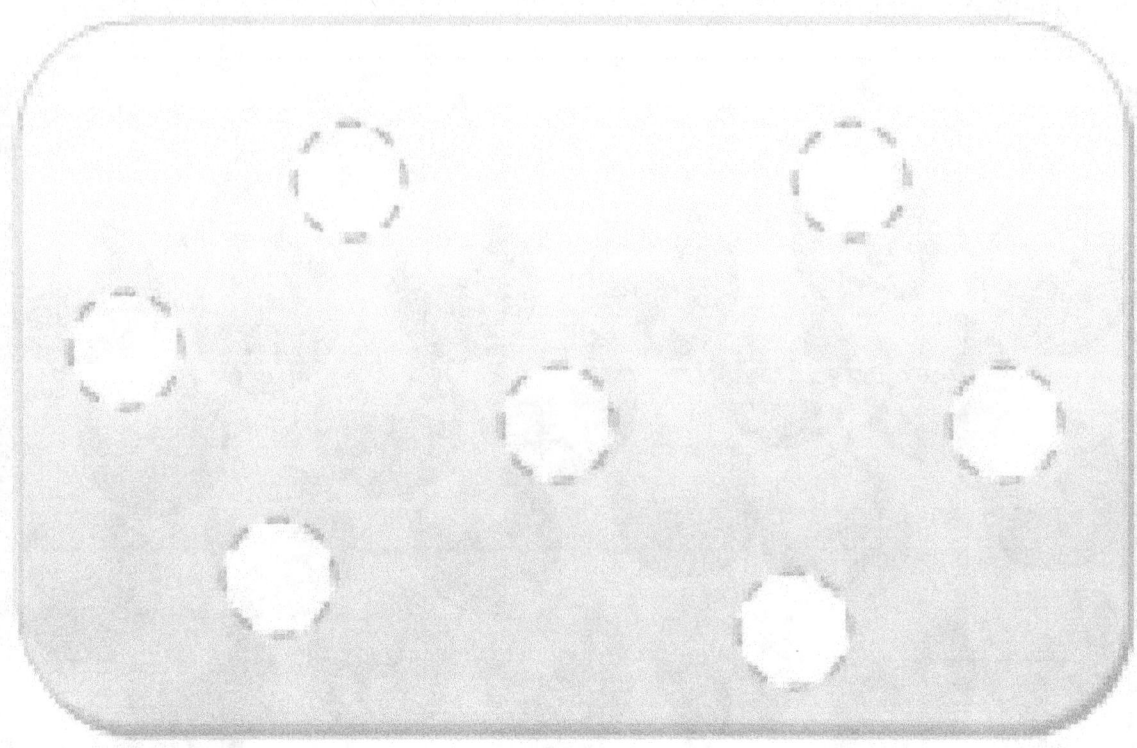

The spaces between gas molecules are very big. Gas molecules have very weak or no bonds at all. The molecules in "gas" can move freely and fast.

without cooling. A vapor can exist in equilibrium with a liquid (or solid), in which case the gas pressure equals the vapor pressure of the liquid (or solid).

A supercritical fluid (SCF) is a gas whose temperature and pressure are above the critical temperature and critical pressure respectively. In this state, the distinction between liquid and gas disappears. A supercritical fluid has the physical properties of a gas, but its high density confers solvent properties in some cases, which leads to useful applications. For example, supercritical carbon dioxide is used to extract caffeine in the manufacture of decaffeinated coffee.[4]

1.1.4 Plasma

Main article: Plasma (physics)

Like a gas, plasma does not have definite shape or volume. Unlike gases, plasmas are electrically conductive, produce magnetic fields and electric currents, and respond strongly to electromagnetic forces. Positively charged nuclei swim in a "sea" of freely-moving disassociated electrons, similar to the way such charges exist in conductive metal. In fact it is this electron "sea" that allows matter in the plasma state to conduct electricity.

The plasma state is often misunderstood, but it is actually quite common on Earth, and the majority of people observe it on a regular basis without even realizing it. Lightning, electric sparks, fluorescent lights, neon lights, plasma televisions, some types of flame and the stars are all examples of illuminated matter in the plasma state.

A gas is usually converted to a plasma in one of two ways, either from a huge voltage difference between two points, or by exposing it to extremely high temperatures.

Heating matter to high temperatures causes electrons to leave the atoms, resulting in the presence of free electrons. At very high temperatures, such as those present in stars, it is assumed that essentially all electrons are "free", and that a very high-energy plasma is essentially bare nuclei swimming in a sea of electrons.

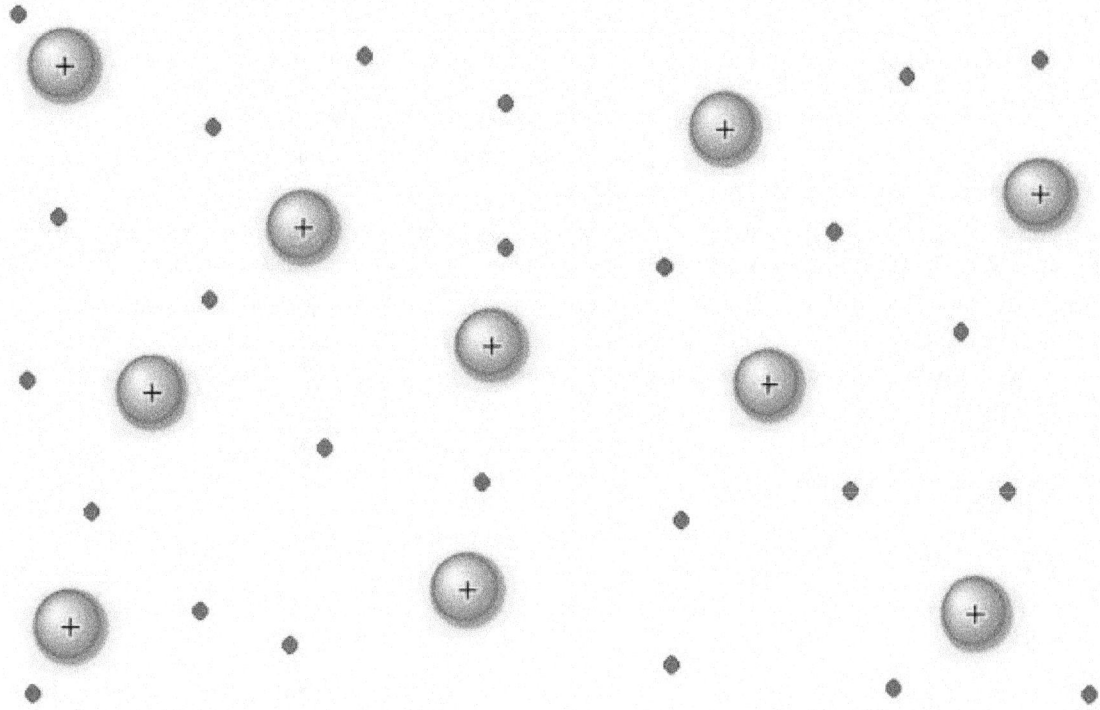

In a plasma, electrons are ripped away from their nuclei, forming an electron "sea". This gives it the ability to conduct electricity.

1.2 Phase transitions

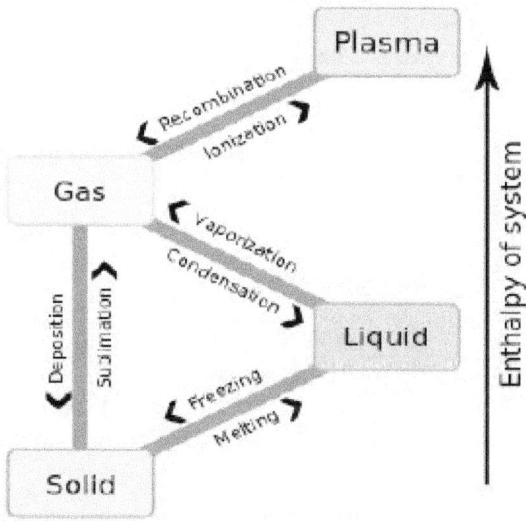

This diagram illustrates transitions between the four fundamental states of matter.

A state of matter is also characterized by phase transitions. A phase transition indicates a change in structure and can be recognized by an abrupt change in properties. A distinct state of matter can be defined as any set of states distinguished from any other set of states by a phase transition. Water can be said to have several distinct solid states.[5] The appearance of superconductivity is associated with a phase transition, so there are superconductive states. Likewise, ferromagnetic states are demarcated by phase transitions and have distinctive properties. When the change of state occurs in stages the intermediate steps are called mesophases. Such phases have been exploited by the introduction of liquid crystal technology. [6][7]

The state or *phase* of a given set of matter can change depending on pressure and temperature conditions, transitioning to other phases as these conditions change to favor their existence; for example, solid transitions to liquid with an increase in temperature. Near absolute zero, a substance exists as a solid. As heat is added to this substance it melts into a liquid at its melting point, boils into a gas at its boiling point, and if heated high enough would enter a plasma state in which the electrons are so energized that they leave their parent atoms.

Forms of matter that are not composed of molecules and are organized by different forces can also be considered different states of matter. Superfluids (like Fermionic condensate) and the quark–gluon plasma are examples.

In a chemical equation, the state of matter of the chemicals may be shown as (s) for solid, (l) for liquid, and (g) for gas. An aqueous solution is denoted (aq). Matter in the plasma state is seldom used (if at all) in chemical equations, so there is no standard symbol to denote it. In the rare equations that plasma is used in plasma is symbolized as (p).

1.3   Non-classical states

1.3.1   Glass

Main article: Glass

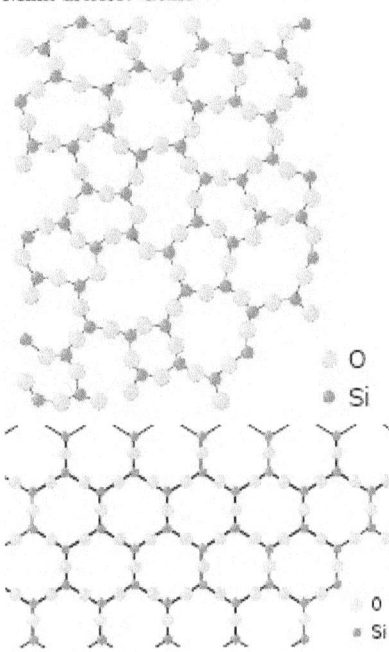

Schematic representation of a random-network glassy form (left) and ordered crystalline lattice (right) of identical chemical composition.

Glass is a non-crystalline or amorphous solid material that exhibits a glass transition when heated towards the liquid state. Glasses can be made of quite different classes of materials: inorganic networks (such as window glass, made of silicate plus additives), metallic alloys, ionic melts, aqueous solutions, molecular liquids, and polymers. Thermodynamically, a glass is in a metastable state with respect to its crystalline counterpart. The conversion rate, however, is practically zero.

1.3.2   Crystals with some degree of disorder

A plastic crystal is a molecular solid with long-range positional order but with constituent molecules retaining rotational freedom; in an orientational glass this degree of freedom is frozen in a quenched disordered state.

Similarly, in a spin glass magnetic disorder is frozen.

1.3.3 Liquid crystal states

Main article: Liquid crystal

Liquid crystal states have properties intermediate between mobile liquids and ordered solids. Generally, they are able to flow like a liquid, but exhibiting long-range order. For example, the nematic phase consists of long rod-like molecules such as para-azoxyanisole, which is nematic in the temperature range 118–136 °C.[8] In this state the molecules flow as in a liquid, but they all point in the same direction (within each domain) and cannot rotate freely.

Other types of liquid crystals are described in the main article on these states. Several types have technological importance, for example, in liquid crystal displays.

1.3.4 Magnetically ordered

Transition metal atoms often have magnetic moments due to the net spin of electrons that remain unpaired and do not form chemical bonds. In some solids the magnetic moments on different atoms are ordered and can form a ferromagnet, an antiferromagnet or a ferrimagnet.

In a ferromagnet—for instance, solid iron—the magnetic moment on each atom is aligned in the same direction (within a magnetic domain). If the domains are also aligned, the solid is a permanent magnet, which is magnetic even in the absence of an external magnetic field. The magnetization disappears when the magnet is heated to the Curie point, which for iron is 768 °C.

An antiferromagnet has two networks of equal and opposite magnetic moments, which cancel each other out so that the net magnetization is zero. For example, in nickel(II) oxide (NiO), half the nickel atoms have moments aligned in one direction and half in the opposite direction.

In a ferrimagnet, the two networks of magnetic moments are opposite but unequal, so that cancellation is incomplete and there is a non-zero net magnetization. An example is magnetite (Fe_3O_4), which contains Fe^{2+} and Fe^{3+} ions with different magnetic moments.

1.3.5 Microphase-separated

Main article: Copolymer

Copolymers can undergo microphase separation to form a diverse array of periodic nanostructures, as shown in the example of the styrene-butadiene-styrene block copolymer shown at right. Microphase separation can be understood by analogy to the phase separation between oil and water. Due to chemical incompatibility between the blocks, block copolymers undergo a similar phase separation. However, because the blocks are covalently bonded to each other, they cannot demix macroscopically as water and oil can, and so instead the blocks form nanometer-sized structures. Depending on the relative lengths of each block and the overall block topology of the polymer, many morphologies can be obtained, each its own phase of matter.

1.3.6 Quantum spin liquid

Main article: Quantum spin liquid

A disordered state in a system of interacting quantum spins which preserves its disorder to very low temperatures, unlike other disordered states.

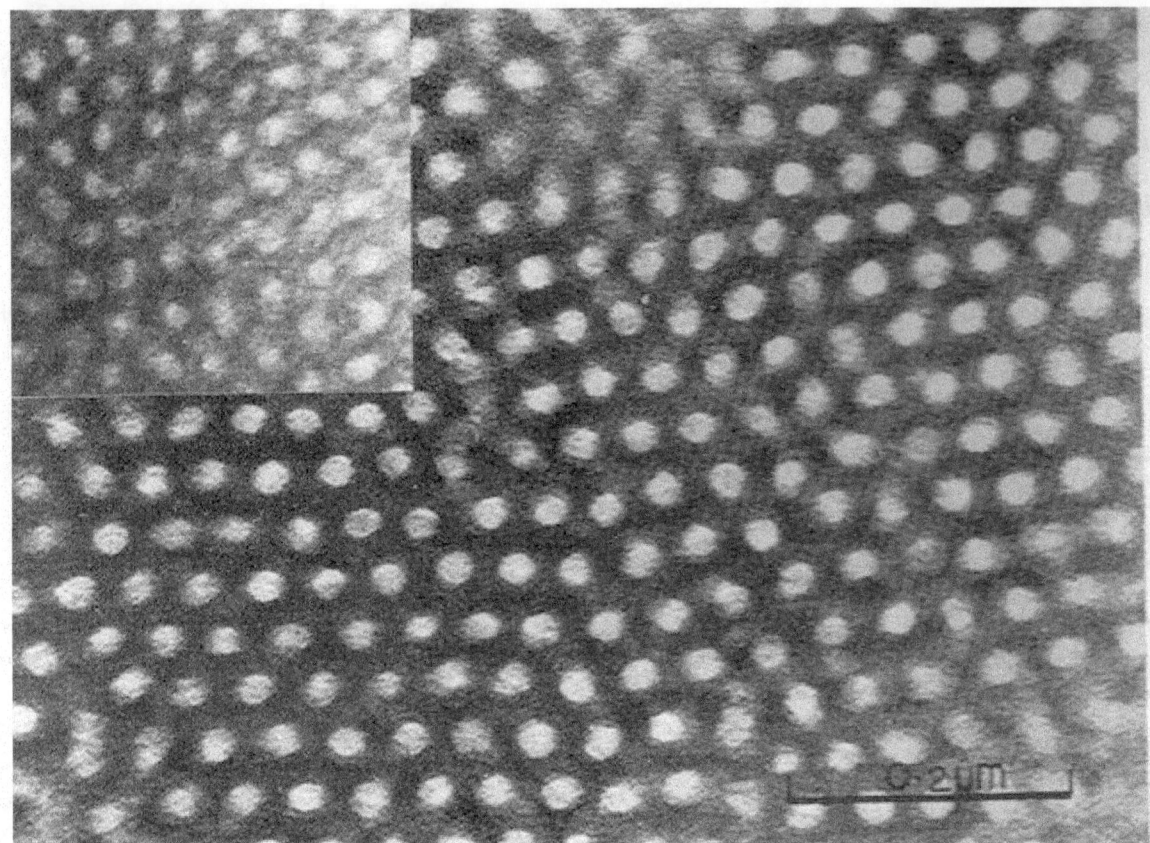

SBS block copolymer in TEM

1.4   Low-temperature states

1.4.1   Superfluid

Main article: Superfluid

Close to absolute zero, some liquids form a second liquid state described as **superfluid** because it has zero viscosity (or infinite fluidity; i.e., flowing without friction). This was discovered in 1937 for helium, which forms a superfluid below the lambda temperature of 2.17 K. In this state it will attempt to "climb" out of its container.[9] It also has infinite thermal conductivity so that no temperature gradient can form in a superfluid. Placing a superfluid in a spinning container will result in quantized vortices.

These properties are explained by the theory that the common isotope helium-4 forms a Bose–Einstein condensate (see next section) in the superfluid state. More recently, Fermionic condensate superfluids have been formed at even lower temperatures by the rare isotope helium-3 and by lithium-6.[10]

1.4.2   Bose–Einstein condensate

Main article: Bose–Einstein condensate

In 1924, Albert Einstein and Satyendra Nath Bose predicted the "Bose–Einstein condensate" (BEC), sometimes referred to as the fifth state of matter. In a BEC, matter stops behaving as independent particles, and collapses into a single quantum state that can be described with a single, uniform wavefunction.

Liquid helium in a superfluid phase creeps up on the walls of the cup in a Rollin film, eventually dripping out from the cup.

In the gas phase, the Bose–Einstein condensate remained an unverified theoretical prediction for many years. In 1995, the research groups of Eric Cornell and Carl Wieman, of JILA at the University of Colorado at Boulder, produced the first such condensate experimentally. A Bose–Einstein condensate is "colder" than a solid. It may occur when atoms have very similar (or the same) quantum levels, at temperatures very close to absolute zero (−273.15 °C).

1.4.3 Fermionic condensate

Main article: Fermionic condensate

A *fermionic condensate* is similar to the Bose–Einstein condensate but composed of fermions. The Pauli exclusion principle prevents fermions from entering the same quantum state, but a pair of fermions can behave as a boson, and multiple such pairs can then enter the same quantum state without restriction.

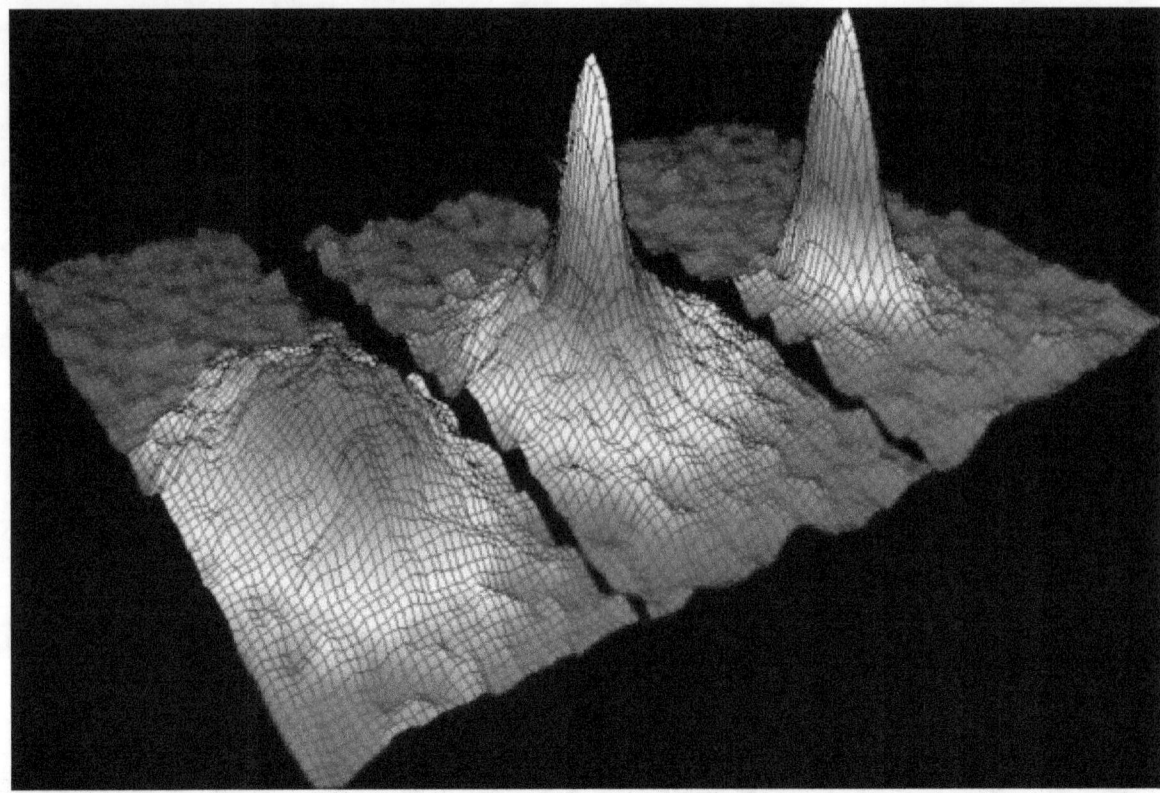

Velocity in a gas of rubidium as it is cooled: the starting material is on the left, and Bose–Einstein condensate is on the right.

1.4.4   Rydberg molecule

One of the metastable states of strongly non-ideal plasma is Rydberg matter, which forms upon condensation of excited atoms. These atoms can also turn into ions and electrons if they reach a certain temperature. In April 2009, *Nature* reported the creation of Rydberg molecules from a Rydberg atom and a ground state atom,[11] confirming that such a state of matter could exist.[12] The experiment was performed using ultracold rubidium atoms.

1.4.5   Quantum Hall state

Main article: Quantum Hall effect

A *quantum Hall state* gives rise to quantized Hall voltage measured in the direction perpendicular to the current flow. A *quantum spin Hall state* is a theoretical phase that may pave the way for the development of electronic devices that dissipate less energy and generate less heat. This is a derivation of the Quantum Hall state of matter.

1.4.6   Strange matter

Main article: Strange matter

Strange matter is a type of quark matter that may exist inside some neutron stars close to the Tolman–Oppenheimer–Volkoff limit (approximately 2–3 solar masses). It may be stable at lower energy states once formed.

1.4.7 Photonic matter

Main article: Photonic matter

In photonic matter, photons behave as if they had mass, and can interact with each other, even forming photonic "molecules". This is in contrast to the usual properties of photons, which have no rest mass, and cannot interact.

1.4.8 Dropleton

Main article: Dropleton

A "quantum fog" of electrons and holes that flow around each other and even ripple like a liquid, rather than existing as discrete pairs.[13]

1.5 High-energy states

1.5.1 Degenerate matter

Main article: Degenerate matter

Under extremely high pressure, ordinary matter undergoes a transition to a series of exotic states of matter collectively known as degenerate matter. In these conditions, the structure of matter is supported by the Pauli exclusion principle. These are of great interest to astrophysicists, because these high-pressure conditions are believed to exist inside stars that have used up their nuclear fusion "fuel", such as the white dwarfs and neutron stars.

Electron-degenerate matter is found inside white dwarf stars. Electrons remain bound to atoms but are able to transfer to adjacent atoms. Neutron-degenerate matter is found in neutron stars. Vast gravitational pressure compresses atoms so strongly that the electrons are forced to combine with protons via inverse beta-decay, resulting in a superdense conglomeration of neutrons. (Normally free neutrons outside an atomic nucleus will decay with a half life of just under 15 minutes, but in a neutron star, as in the nucleus of an atom, other effects stabilize the neutrons.)

1.5.2 Quark–gluon plasma

Main article: Quark–gluon plasma

Quark–gluon plasma is a phase in which quarks become free and able to move independently (rather than being perpetually bound into particles) in a sea of gluons (subatomic particles that transmit the strong force that binds quarks together); this is similar to splitting molecules into atoms. This state may be briefly attainable in particle accelerators, and allows scientists to observe the properties of individual quarks, and not just theorize. See also Strangeness production.

Quark–gluon plasma was discovered at CERN in 2000.

1.5.3 Color-glass condensate

Main article: Color-glass condensate

Color-glass condensate is a type of matter theorized to exist in atomic nuclei traveling near the speed of light. According to Einstein's theory of relativity, a high-energy nucleus appears length contracted, or compressed, along its direction of motion. As a result, the gluons inside the nucleus appear to a stationary observer as a "gluonic wall" traveling near the speed of light. At very high energies, the density of the gluons in this wall is seen to increase greatly. Unlike the quark–gluon plasma produced in the collision of such walls, the color-glass condensate describes the walls themselves, and is an intrinsic property of the particles that can only be observed under high-energy conditions such as those at RHIC and possibly at the Large Hadron Collider as well.

1.6   Very high energy states

The gravitational singularity predicted by general relativity to exist at the center of a black hole is *not* a phase of matter; it is not a material object at all (although the mass-energy of matter contributed to its creation) but rather a property of spacetime at a location. It could be argued, of course, that all particles are properties of spacetime at a location,[14] leaving a half-note of controversy on the subject.

1.7   Other proposed states

1.7.1   Supersolid

Main article: Supersolid

A supersolid is a spatially ordered material (that is, a solid or crystal) with superfluid properties. Similar to a superfluid, a supersolid is able to move without friction but retains a rigid shape. Although a supersolid is a solid, it exhibits so many characteristic properties different from other solids that many argue it is another state of matter.[15]

1.7.2   String-net liquid

Main article: String-net liquid

In a string-net liquid, atoms have apparently unstable arrangement, like a liquid, but are still consistent in overall pattern, like a solid. When in a normal solid state, the atoms of matter align themselves in a grid pattern, so that the spin of any electron is the opposite of the spin of all electrons touching it. But in a string-net liquid, atoms are arranged in some pattern that requires some electrons to have neighbors with the same spin. This gives rise to curious properties, as well as supporting some unusual proposals about the fundamental conditions of the universe itself.

1.7.3   Superglass

Main article: Superglass

A superglass is a phase of matter characterized, at the same time, by superfluidity and a frozen amorphous structure.

1.7.4   Dark matter

Main article: Dark matter

While dark matter is estimated to comprise 83% of the mass of matter in the universe, most of its properties remain a mystery due to the fact that it neither absorbs nor emits electromagnetic radiation, and there are many competing theories regarding what dark matter is actually made of. Thus, while it is hypothesized to exist and comprise the vast majority of matter in the universe, almost all of its properties are unknown and a matter of speculation, because it has only been observed through its gravitational effects.[16][17]

1.7.5 Equilibrium gel

Main article: Equilibrium gel

Equilibrium gel is made from a synthetic clay called Laponite. Unlike other gels, it maintains the same consistency throughout its structure and is stable, which means it does not separate into sections of solid mass and those of more liquid mass. Equilibrium gel filtration liquid chromatography is a technique used for the quantitation of ligand binding.[18]

1.8 See also

- Hidden states of matter

- Classical element

- Condensed matter physics

- Cooling curve

- Phase (matter)

- Supercooling

- Superheating

1.9 Notes and references

[1] It is often stated that more than 99% of the material in the visible universe is plasma. See, for instance, D. A. Gurnett; A. Bhattacharjee (2005). *Introduction to Plasma Physics: With Space and Laboratory Applications.* Cambridge, UK: Cambridge University Press. p. 2. ISBN 0-521-36483-3. and K Scherer; H Fichtner; B Heber (2005). *Space Weather: The Physics Behind a Slogan.* Berlin: Springer. p. 138. ISBN 3-540-22907-8.. Essentially, all of the visible light from space comes from stars, which are plasmas with a temperature such that they radiate strongly at visible wavelengths. Most of the ordinary (or baryonic) matter in the universe, however, is found in the intergalactic medium, which is also a plasma, but much hotter, so that it radiates primarily as X-rays. The current scientific consensus is that about 96% of the total energy density in the universe is not plasma or any other form of ordinary matter, but a combination of cold dark matter and dark energy.

[2] M.A. Wahab (2005). *Solid State Physics: Structure and Properties of Materials.* Alpha Science. pp. 1–3. ISBN 1-84265-218-4.

[3] F. White (2003). *Fluid Mechanics.* McGraw-Hill. p. 4. ISBN 0-07-240217-2.

[4] G. Turrell (1997). *Gas Dynamics: Theory and Applications.* John Wiley & Sons. pp. 3–5. ISBN 0-471-97573-7.

[5] M. Chaplin (20 August 2009). "Water phase Diagram". *Water Structure and Science.* Retrieved 23 February 2010.

[6] D.L. Goodstein (1985). *States of Matter.* Dover Phoenix. ISBN 978-0-486-49506-4.

[7] A.P. Sutton (1993). *Electronic Structure of Materials.* Oxford Science Publications. pp. 10–12. ISBN 978-0-19-851754-2.

[8] Shao, Y.; Zerda, T. W. (1998). "Phase Transitions of Liquid Crystal PAA in Confined Geometries". *Journal of Physical Chemistry B* 102 (18): 3387–3394. doi:10.1021/jp9734437.

[9] J.R. Minkel (20 February 2009). "Strange but True: Superfluid Helium Can Climb Walls". *Scientific American.* Retrieved 23 February 2010.

[10] L. Valigra (22 June 2005). "MIT physicists create new form of matter". MIT News. Retrieved 23 February 2010.

[11] V. Bendkowsky; et al. (2009). "Observation of Ultralong-Range Rydberg Molecules". *Nature* 458 (7241): 1005–doi:10.1038/nature07945. PMID 19396141.

[12] V. Gill (23 April 2009). "World First for Strange Molecule". BBC News. Retrieved 23 February 2010.

[13] http://www.iflscience.com/physics/new-state-matter-discovered#3Oe9x65kkHViXABt.99

[14] David Chalmers; David Manley; Ryan Wasserman (2009). *Metametaphysics: New Essays on the Foundations of Ontology*. Oxford University Press. pp. 378–. ISBN 978-0-19-954604-6.

[15] G. Murthy; et al. (1997). "Superfluids and Supersolids on Frustrated Two-Dimensional Lattices". *Physical Review B* 55 (5): 3104. arXiv:cond-mat/9607217. Bibcode:1997PhRvB..55.3104M. doi:10.1103/PhysRevB.55.3104.

[16] Trimble, Virginia (1987). "Existence and nature of dark matter in the universe". *Annual Review of Astronomy and Astrophysics* 25: 425–472. Bibcode:1987ARA&A..25..425T. doi:10.1146/annurev.aa.25.090187.002233.

[17] Hinshaw, Gary F. (29 January 2010). "What is the universe made of?". *Universe 101*. NASA website. Retrieved 17 March 2010.

[18] Cartlidge, Edwin (12 January 2012). "New State of Matter Seen in Clay". *Technology*. Science Now website. Retrieved 10 September 2013.

1.10   External links

- 2005-06-22, MIT News: MIT physicists create new form of matter Citat: "... They have become the first to create a new type of matter, a gas of atoms that shows high-temperature superfluidity."

- 2003-10-10, Science Daily: Metallic Phase For Bosons Implies New State Of Matter

- 2004-01-15, ScienceDaily: Probable Discovery Of A New, Supersolid, Phase Of Matter Citat: "...We apparently have observed, for the first time, a solid material with the characteristics of a superfluid...but because all its particles are in the identical quantum state, it remains a solid even though its component particles are continually flowing..."

- 2004-01-29, ScienceDaily: NIST/University Of Colorado Scientists Create New Form Of Matter: A Fermionic Condensate

- Short videos demonstrating of States of Matter, solids, liquids and gases by Prof. J M Murrell, University of Sussex

Chapter 2

Enthalpy of fusion

For the plastic welding technique, see Heat fusion.

The **enthalpy of fusion** also known as **(latent) heat of fusion** is the change in enthalpy resulting from heating a given

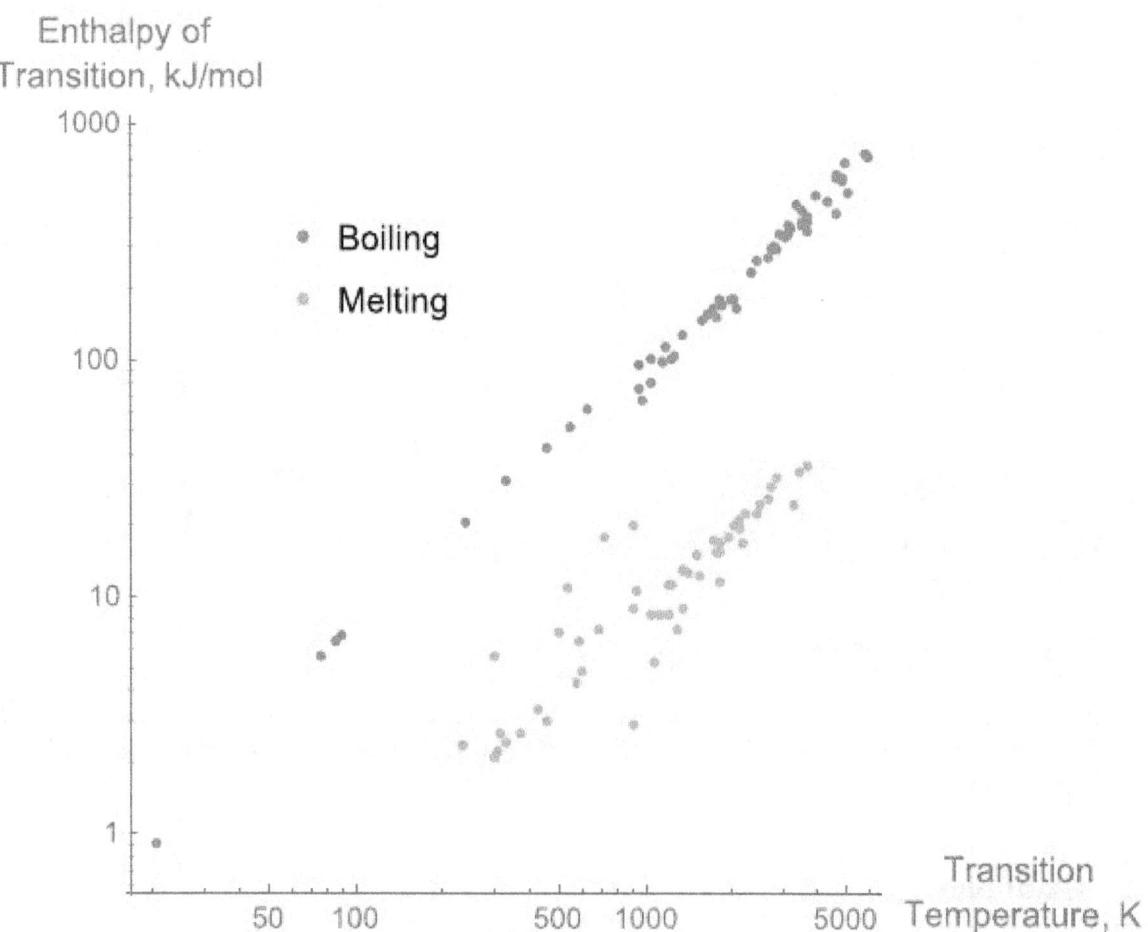

Enthalpies of melting and boiling for pure elements versus temperatures of transition, demonstrating Trouton's rule.

quantity of a substance to change its state from a solid to a liquid. The temperature at which this occurs is the melting point.

The 'enthalpy' of fusion is a latent heat, because during melting the introduction of heat cannot be observed as a temper-

15

ature change, as the temperature remains constant during the process. The latent heat of fusion is the enthalpy change of any amount of substance when it melts. When the heat of fusion is referenced to a unit of mass, it is usually called the **specific heat of fusion**, while the **molar heat of fusion** refers to the enthalpy change per amount of substance in moles.

The liquid phase has a higher internal energy than the solid phase. This means energy must be supplied to a solid in order to melt it and energy is released from a liquid when it freezes, because the molecules in the liquid experience weaker intermolecular forces and so have a higher potential energy (a kind of bond-dissociation energy for intermolecular forces).

When liquid water is cooled, its temperature falls steadily until it drops just below the line of freezing point at 0 °C. The temperature then remains constant at the freezing point while the water crystallizes. Once the water is completely frozen, its temperature continues to fall.

The enthalpy of fusion is almost always a positive quantity; helium is the only known exception.[1] Helium-3 has a negative enthalpy of fusion at temperatures below 0.3 K. Helium-4 also has a very slightly negative enthalpy of fusion below 0.8 K. This means that, at appropriate constant pressures, these substances freeze with the addition of heat.[2]

2.1   Reference values of common substances

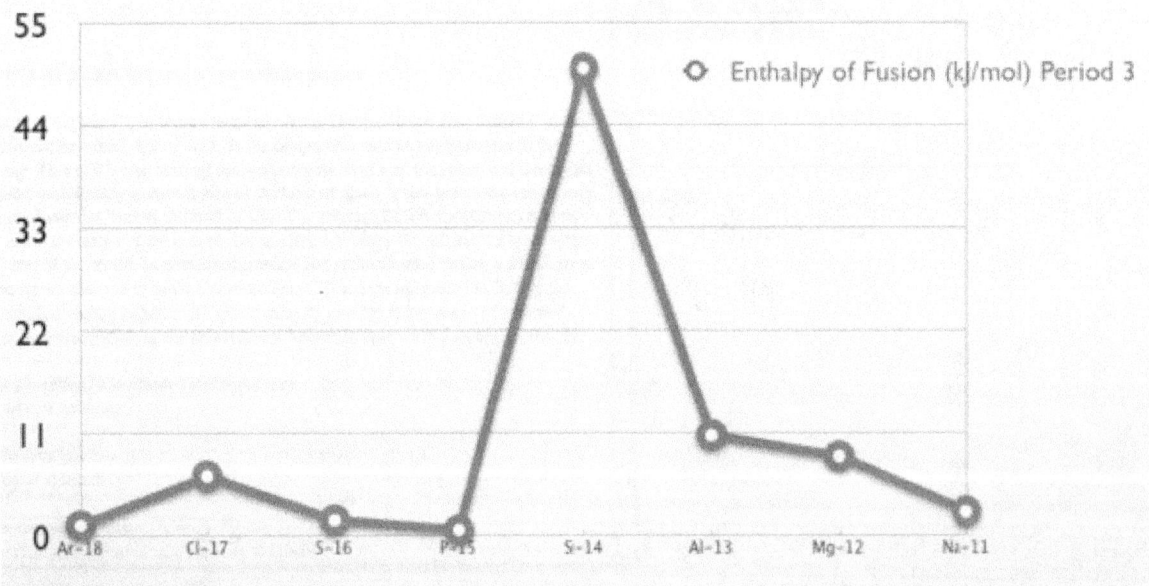

Standard enthalpy change of fusion of period three.

These values are from the CRC *Handbook of Chemistry and Physics*, 62nd edition. The conversion between cal/g and J/g in the above table uses the thermochemical calorie $(cal_{th}) = 4.184$ joules rather than the International Steam Table calorie (calINT) = 4.1868 joules.

2.2   Examples

1) To heat 1 kg (about 1 liter) of water from 283.15 K to 303.15 K (10 °C to 30 °C) requires 83.6 kJ. However, to melt ice also requires energy. To heat ice from 273.15 K to water at 293.15 K (0 °C to 20 °C) requires:

   (1) 333.55 J/g (heat of fusion of ice) = 333.55 kJ/kg = 333.55 kJ for 1 kg of ice to melt

   PLUS

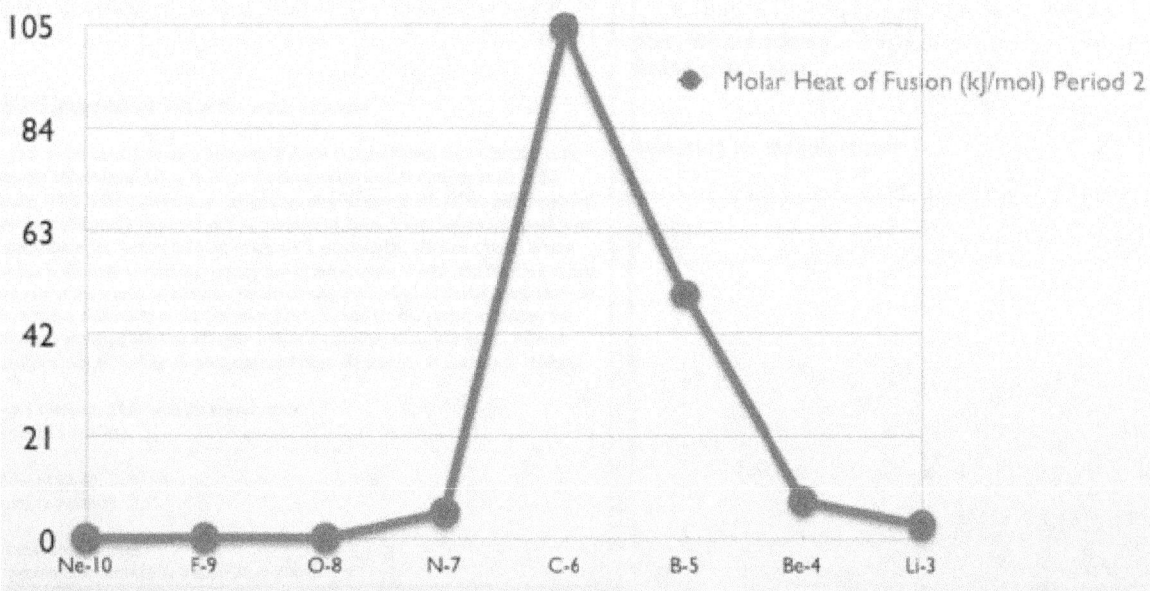

Standard enthalpy change of fusion of period two of the periodic table of elements.

(2) 4.18 J/(g·K) × 20K = 4.18 kJ/(kg·K) × 20K = 83.6 kJ for 1 kg of water to increase in temperature by 20 K

= 417.15 kJ

Thus one part ice at 0 °C will cool almost exactly 4 parts water from 20 °C to 0 °C.

2) Silicon has a heat of fusion of 50.21 kJ/mol. 50 kW of power can supply the energy required to melt about 100 kg of silicon in one hour, after it is brought to the melting point temperature:

50 kW = 50kJ/s = 180000kJ/h

180000kJ/h * (1 mol Si)/50.21kJ * 28gSi/(mol Si) * 1kgSi/1000gSi = 100.4kg/h

2.3 Solubility prediction

The heat of fusion can also be used to predict solubility for solids in liquids. Provided an ideal solution is obtained the mole fraction (x_2) of solute at saturation is a function of the heat of fusion, the melting point of the solid (T_{fus}) and the temperature (T) of the solution:

$$\ln x_2 = -\frac{\Delta H^o_{fus}}{R}\left(\frac{1}{T} - \frac{1}{T_{fus}}\right)$$

Here, R is the gas constant. For example the solubility of paracetamol in water at 298 K is predicted to be:

$$x_2 = \exp\left(-\frac{28100 \text{ J mol}^{-1}}{8.314 \text{ J K}^{-1} \text{ mol}^{-1}}\left(\frac{1}{298} - \frac{1}{442}\right)\right) = 0.0248$$

This equals to a solubility in grams per liter of:

$$\frac{0.0248 * \frac{1000 \frac{\text{g}}{\text{mol}}}{18.053}}{1 - 0.0248} * 151.17 \text{ mol}^{-1} = 213.4$$

which is a deviation from the real solubility (240 g/L) of 11%. This error can be reduced when an additional heat capacity parameter is taken into account.[3]

2.3.1   Proof

At equilibrium the chemical potentials for the pure solvent and pure solid are identical:

$$\mu^{\circ}_{solid} = \mu^{\circ}_{solution}$$

or

$$\mu^{\circ}_{solid} = \mu^{\circ}_{liquid} + RT \ln X_2$$

with R the gas constant and T the temperature.

Rearranging gives:

$$RT \ln X_2 = -(\mu^{\circ}_{liquid} - \mu^{\circ}_{solid})$$

and since

$$\Delta G^{\circ}_{fus} = -(\mu^{\circ}_{liquid} - \mu^{\circ}_{solid})$$

the heat of fusion being the difference in chemical potential between the pure liquid and the pure solid, it follows that

$$RT \ln X_2 = -(\Delta G^{\circ}_{fus})$$

Application of the Gibbs–Helmholtz equation:

$$\left(\frac{\partial(\frac{\Delta G^{\circ}_{fus}}{T})}{\partial T} \right)_P = - \frac{\Delta H^{\circ}_{fus}}{T^2}$$

ultimately gives:

$$\left(\frac{\partial(\ln X_2)}{\partial T} \right) = \frac{\Delta H^{\circ}_{fus}}{RT^2}$$

or:

$$\partial \ln X_2 = \frac{\Delta H^{\circ}_{fus}}{RT^2} * \delta T$$

and with integration:

$$\int_{x_2=1}^{x_2=x_2} \delta \ln X_2 = \ln x_2 = \int_{T_{fus}}^{T} \frac{\Delta H^{\circ}_{fus}}{RT^2} * \Delta T$$

the end result is obtained:

$$\ln x_2 = - \frac{\Delta H^{\circ}_{fus}}{R} \left(\frac{1}{T} - \frac{1}{T_{fus}} \right)$$

2.4   See also

- Heat of vaporization

- Heat capacity

- Thermodynamic databases for pure substances

- Joback method (Estimation of the heat of fusion from molecular structure)

- Latent heat

2.5   Notes

[1] Atkins & Jones 2008, p. 236.

[2] Ott & Boerio-Goates 2000, pp. 92–93.

[3] *Measurement and Prediction of Solubility of Paracetamol in Water-Isopropanol Solution. Part 2. Prediction* H. Hojjati and S. Rohani Org. Process Res. Dev.; **2006**; 10(6) pp 1110–1118; (Article) doi:10.1021/op060074g

2.6   References

- Atkins, Peter; Jones, Loretta (2008), *Chemical Principles: The Quest for Insight* (4th ed.), W. H. Freeman and Company, p. 236, ISBN 0-7167-7355-4

- Ott, J. Bevan; Boerio-Goates, Juliana (2000), *Chemical Thermodynamics: Advanced Applications*, Academic Press, ISBN 0-12-530985-6

Chapter 3

Enthalpy of sublimation

The **enthalpy of sublimation**, or **heat of sublimation**, is the heat required to sublime one mole of the substance at a given combination of temperature and pressure, usually standard temperature and pressure (STP). The heat of sublimation is usually expressed in kJ/mol, although the less customary kJ/kg is also encountered.

3.1   Sublimation enthalpy substances

3.2   See also

- Heat
- Sublimation (chemistry)
- Phase transition

Chapter 4

Enthalpy of vaporization

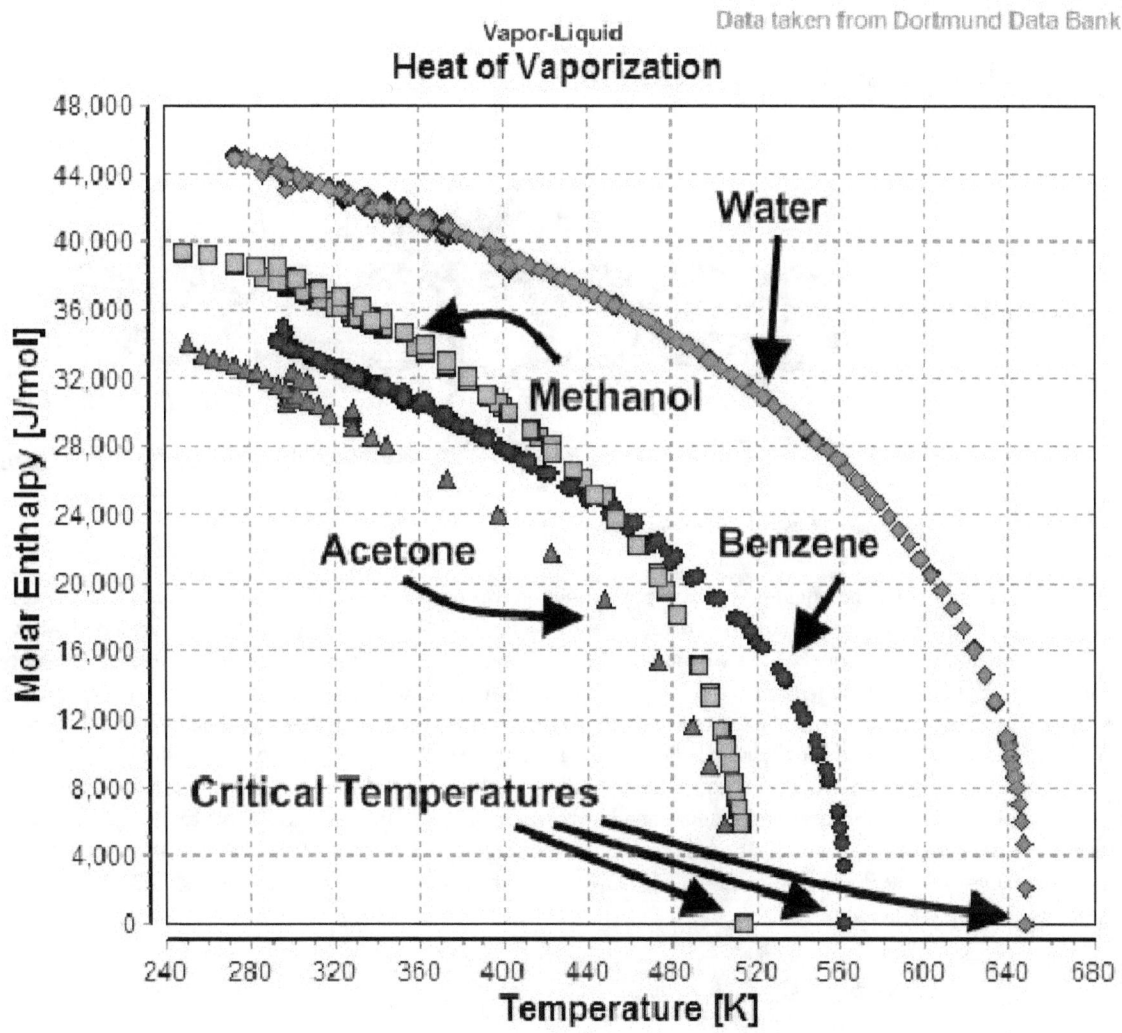

Temperature-dependency of the heats of vaporization for water, methanol, benzene, and acetone.

The **enthalpy of vaporization**, (symbol ΔH_{vap}) also known as the **(latent) heat of vaporization** or **heat of evaporation**, is the enthalpy change required to transform a given quantity of a substance from a liquid into a gas at a given pressure (often atmospheric pressure, as in STP).

It is often measured at the normal boiling point of a substance; although tabulated values are usually corrected to 298 K, the correction is often smaller than the uncertainty in the measured value.

The heat of vaporization is temperature-dependent, though a constant heat of vaporization can be assumed for small temperature ranges and for reduced temperature $T_r \ll 1$. The heat of vaporization diminishes with increasing temperature and it vanishes completely at the critical temperature ($T_r = 1$) because above the critical temperature the liquid and vapor phases no longer exist, since the substance is a supercritical fluid.

4.1   Units

Values are usually quoted in J/mol or kJ/mol (molar enthalpy of vaporization), although kJ/kg or J/g (specific heat of vaporization), and older units like kcal/mol, cal/g and Btu/lb are sometimes still used, among others.

4.2   Physical model for vaporization

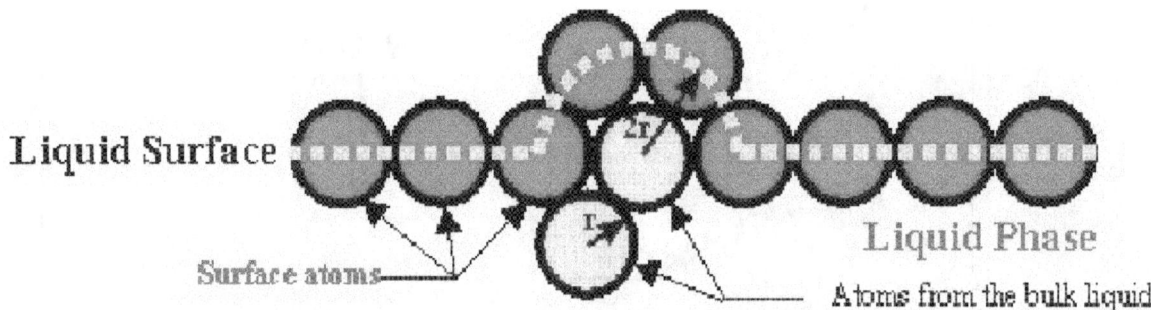

Fig. 1 Schematic cross section of the proposed vaporization model for monatomic liquids with one atomic surface layer.

A simple physical model for the liquid–gas phase transformation was proposed in 2009 by Jozsef Garai.[1] It is suggested that the energy required to free an atom from the liquid is equivalent to the energy needed to overcome the surface resistance of the liquid. The model allows calculating the latent heat by multiplying the maximum surface area covering an atom (Fig. 1) with the surface tension and the number of atoms in the liquid. The calculated latent heat of vaporization values for the investigated 45 elements agrees well with experiments. Another model which utilizes the data set from Jozsef Garai's model shows that the liquid–gas phase change can be explained in terms of kinetic theory by considering that the energy required for vaporization is extracted from all six of the vaporizing molecule's neighbours. This includes a required rethink of the probability of vaporization, and has consequences to the Clausius-Clapeyron equation. Moreover, it does resolve the issue of the latent heat of vaporization being significantly greater than the thermal energy exchanged between molecules, i.e. at boiling point the latent heat for water is approximately 13.2 times kT (Boltzmann's factor multiplied by boiling temperature.)[2]

4.3   Enthalpy of condensation

The **enthalpy of condensation** (or **heat of condensation**) is by definition equal to the enthalpy of vaporization with the opposite sign: enthalpy changes of vaporization are always positive (heat is absorbed by the substance), whereas enthalpy changes of condensation are always negative (heat is released by the substance).

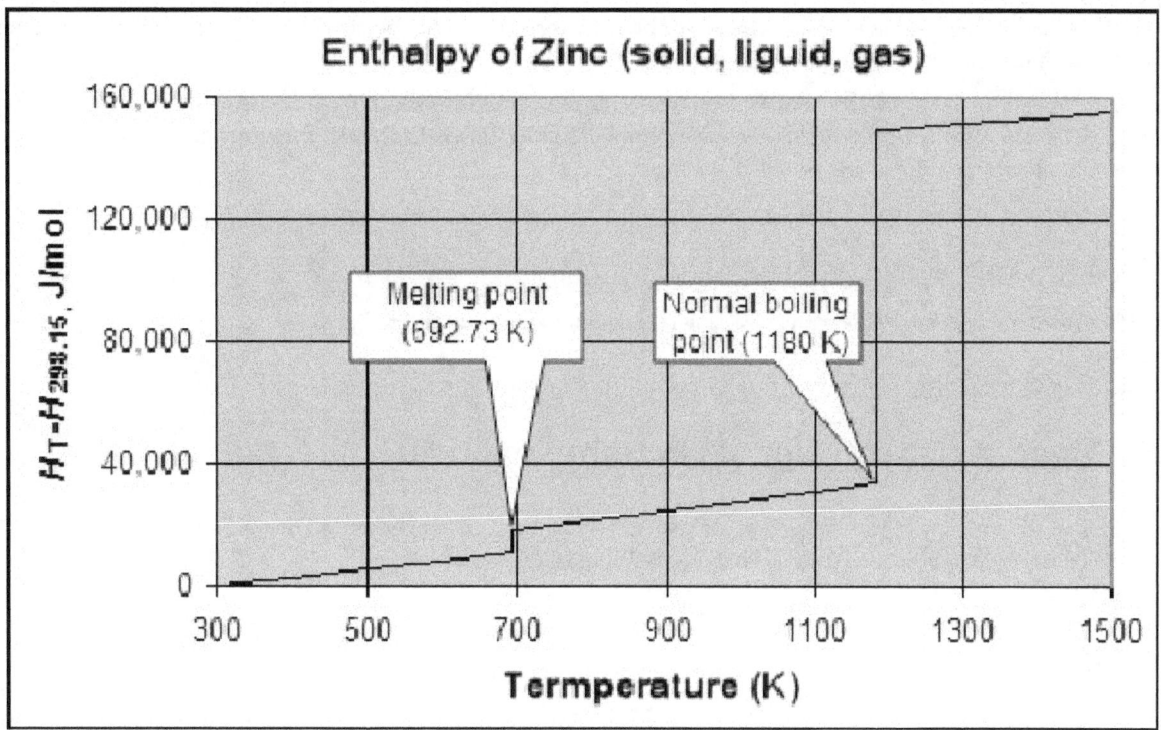

Molar enthalpy of zinc above 298.15 K and at 1 atm pressure, showing discontinuities at the melting and boiling points. The enthalpy of melting ($\Delta H°m$) of zinc is 7323 J/mol, and the enthalpy of vaporization ($\Delta H°v$) is 115 330 J/mol.

4.4 Thermodynamic background

The enthalpy of vaporization can be written as

$$\Delta H_{vap} = \Delta U_{vap} + p\Delta V$$

It is equal to the increased internal energy of the vapor phase compared with the liquid phase, plus the work done against ambient pressure. The increase in the internal energy can be viewed as the energy required to overcome the intermolecular interactions in the liquid (or solid, in the case of sublimation). Hence helium has a particularly low enthalpy of vaporization, 0.0845 kJ/mol, as the van der Waals forces between helium atoms are particularly weak. On the other hand, the molecules in liquid water are held together by relatively strong hydrogen bonds, and its enthalpy of vaporization, 40.65 kJ/mol, is more than five times the energy required to heat the same quantity of water from 0 °C to 100 °C ($c_p = 75.3$ J K^{-1} mol^{-1}). Care must be taken, however, when using enthalpies of vaporization to *measure* the strength of intermolecular forces, as these forces may persist to an extent in the gas phase (as is the case with hydrogen fluoride), and so the calculated value of the bond strength will be too low. This is particularly true of metals, which often form covalently bonded molecules in the gas phase: in these cases, the enthalpy of atomization must be used to obtain a true value of the bond energy.

An alternative description is to view the enthalpy of condensation as the heat which must be released to the surroundings to compensate for the drop in entropy when a gas condenses to a liquid. As the liquid and gas are in equilibrium at the boiling point (T_b), $\Delta_v G = 0$, which leads to:

$$\Delta_v S = S_{gas} - S_{liquid} = \Delta_v H / T_b$$

As neither entropy nor enthalpy vary greatly with temperature, it is normal to use the tabulated standard values without any correction for the difference in temperature from 298 K. A correction must be made if the pressure is different from

100 kPa, as the entropy of a gas is proportional to its pressure (or, more precisely, to its fugacity): the entropies of liquids vary little with pressure, as the compressibility of a liquid is small.

These two definitions are equivalent: the boiling point is the temperature at which the increased entropy of the gas phase overcomes the intermolecular forces. As a given quantity of matter always has a higher entropy in the gas phase than in a condensed phase ($\Delta_v S$ is always positive), and from

$$\Delta G = \Delta H - T \Delta S$$

the Gibbs free energy change falls with increasing temperature: gases are favored at higher temperatures, as is observed in practice.

4.5 Vaporization enthalpy of electrolyte solutions

Estimation of the enthalpy of vaporization of electrolyte solutions can be simply carried out using equations based on the chemical thermodynamic models, such as Pitzer model[3] or TCPC model.[4]

4.6 Selected values

4.6.1 Elements

4.6.2 Other common substances

Enthalpies of vaporization of common substances, measured at their respective standard boiling points:

4.7 See also

- Enthalpy of fusion

- Enthalpy of sublimation

- Joback method (Estimation of the heat of vaporization at the normal boiling point from molecular structures)

4.8 References

[1] Garai, J. (2009). "Physical model for vaporization". *Fluid Phase Equilibria* **283**: 89–77. doi:10.1016/j.fluid.2009.06.005.

[2] K. Mayhew, Phys. Essays 26, vol 4, 604 (2013)

[3] X. Ge, X. Wang. Estimation of Freezing Point Depression, Boiling Point Elevation and Vaporization enthalpies of electrolyte solutions. Ind. Eng. Chem. Res. 48(2009)2229–2235. http://pubs.acs.org/doi/abs/10.1021/ie801348c (Correction: 2009, 48, 5123)http://pubs.acs.org/doi/abs/10.1021/ie900434h

[4] X. Ge, X. Wang. Calculations of Freezing Point Depression, Boiling Point Elevation, Vapor Pressure and Enthalpies of Vaporization of Electrolyte Solutions by a Modified Three-Characteristic Parameter Correlation Model. J. Sol. Chem. 38(2009)1097–1117.http://www.springerlink.com/content/21670685448p5145/

- CODATA Key Values for Thermodynamics

- Kugler HK & Keller C (eds) 1985, *Gmelin handbook of inorganic and organometallic chemistry*, 8th ed., 'At, Astatine', system no. 8a, Springer-Verlag, Berlin, ISBN 3-540-93516-9, pp. 116–117

- NIST Chemistry WebBook

- Sears, Zemansky et al., *University Physics*, Addison-Wesley Publishing Company, Sixth ed., 1982, ISBN 0-201-07199-1

Chapter 5

Latent heat

Latent heat is energy released or absorbed, by a body or a thermodynamic system, during a constant-temperature process that is specified in some way. An example is latent heat of fusion for a phase change, melting, at a specified temperature and pressure.[1][2] The term was introduced around 1762 by Scottish chemist Joseph Black. It is derived from the Latin *latere* (*to lie hidden*). Black used the term in the context of calorimetry where a heat transfer caused a volume change while the thermodynamic system's temperature was constant.

In contrast to latent heat, sensible heat involves an energy transfer that results in a temperature change of the system.

5.1   Usage

The terms "sensible heat" and "latent heat" are not special forms of energy; instead they measure two kinds of change in a material or in a thermodynamic system. "Sensible heat" measures change in a body's internal energy that may be "sensed" with a thermometer. "Latent heat" measures change in internal energy that seems hidden from a thermometer – the temperature reading doesn't change. Heat is energy in the process of transferring between a system and its surroundings, other than as work or by transfer of matter.

Both sensible and latent heats are observed in many processes of transport of energy in nature. Latent heat is associated with the phase changes of atmospheric water vapor, mostly vaporization and condensation, whereas sensible heat is energy transferred that affects the temperature of the atmosphere.

The original usage of the term, as introduced by Black, was applied to systems that were intentionally held at constant temperature. Such usage referred to *latent heat of expansion* and several other related latent heats. These latent heats are defined independently of the conceptual framework of thermodynamics.[3]

When a body is heated at constant temperature by thermal radiation in a microwave field for example, it may expand by an amount described by its *latent heat with respect to volume* or *latent heat of expansion*, or increase its pressure by an amount described by its *latent heat with respect to pressure*.[4]

Two common forms of latent heat are latent heat of fusion (melting) and latent heat of vaporization (boiling). These names describe the direction of energy flow when changing from one phase to the next: from solid to liquid, and liquid to gas.

In both cases the change is endothermic, meaning that the system absorbs energy. If the change is exothermic, then energy is released. For example, when water evaporates, energy is transferred from a water molecule to an air molecule that contains less water vapor than its surroundings. Because energy is required for the water molecule to overcome the forces of attraction between water particles, the transition from water to vapor requires an input of energy and causes a temperature drop in the water molecule's surroundings.

If the vapor then condenses to a liquid on a surface, then the vapor's latent energy absorbed during evaporation is released as the liquid's sensible heat onto the surface.

26

The large value of the enthalpy of condensation of water vapor is the reason that steam is a far more effective heating medium than boiling water, and is more hazardous.

5.1.1 Meteorology

In meteorology, latent heat flux is the flux of heat from the Earth's surface to the atmosphere that is associated with evaporation or transpiration of water at the surface and subsequent condensation of water vapor in the troposphere. It is an important component of Earth's surface energy budget. Latent heat flux has been commonly measured with the Bowen ratio technique, or more recently since the mid-1900s by the eddy covariance method.

5.2 History

The English word *latent* comes from Latin *latēns*, meaning *lying hidden*.[5][6] The term *latent heat* was introduced into calorimetry around 1750 when Joseph Black, commissioned by producers of Scotch whisky in search of ideal quantities of fuel and water for their distilling process,[7] to studying system changes, such as of volume and pressure, when the thermodynamic system was held at constant temperature in a thermal bath. James Prescott Joule characterised latent energy as the energy of interaction in a given configuration of particles, i.e. a form of potential energy, and the sensible heat as an energy that was indicated by the thermometer,[8] relating the latter to thermal energy.

5.3 Specific latent heat

A *specific* latent heat (L) expresses the amount of energy in the form of heat (Q) required to completely effect a phase change of a unit of mass (m), usually 1kg, of a substance as an intensive property:

$$L = \frac{Q}{m}.$$

Intensive properties are material characteristics and are not dependent on the size or extent of the sample. Commonly quoted and tabulated in the literature are the specific latent heat of fusion and the specific latent heat of vaporization for many substances.

From this definition, the latent heat for a given mass of a substance is calculated by

$$Q = mL$$

where:

> Q is the amount of energy released or absorbed during the change of phase of the substance (in kJ or in BTU),
>
> m is the mass of the substance (in kg or in lb), and
>
> L is the specific latent heat for a particular substance (kJ-kg_m^{-1} or in BTU-lb_m^{-1}), either L_f for fusion, or L_v for vaporization.

5.4 Table of latent heats

The following table shows the latent heats and change of phase temperatures of some common fluids and gases.

5.5   Latent heat for condensation of water

The latent heat of condensation of water in the temperature range from −25 °C to 40 °C is approximated by the following empirical cubic function:

$$L_{\text{water}}(T) = (2500.8 - 2.36T + 0.0016T^2 - 0.00006T^3) \text{ J/g}, \text{ [10]}$$

where the temperature T is taken to be the numerical value in °C.

For sublimation and deposition from and into ice, the latent heat is almost constant in the temperature range from −40 °C to 0 °C and can be approximated by the following empirical quadratic function:

$$L_{\text{ice}}(T) = (2834.1 - 0.29T - 0.004T^2) \text{ J/g. [10]}$$

5.6   See also

- Bowen ratio

- Eddy covariance flux (eddy correlation, eddy flux)

- Sublimation (physics)

- Specific heat capacity

- Enthalpy of fusion

- Enthalpy of vaporization

5.7   References

[1] Perrot, Pierre (1998). *A to Z of Thermodynamics*. Oxford University Press. ISBN 0-19-856552-6.

[2] Clark, John, O.E. (2004). *The Essential Dictionary of Science*. Barnes & Noble Books. ISBN 0-7607-4616-8.

[3] Bryan, G.H. (1907). *Thermodynamics. An Introductory Treatise dealing mainly with First Principles and their Direct Applications*, B.G. Tuebner, Leipzig, pages 9, 20–22.

[4] Maxwell, J.C. (1872). *Theory of Heat*, third edition, Longmans, Green, and Co., London, page 73.

[5] Harper, Douglas. "latent". *Online Etymology Dictionary*.

[6] Lewis, Charlton T. (1890). *An Elementary Latin Dictionary*. Entry for latens.

[7] James Burke (1979). "Credit Where It's Due". *The Day the Universe Changed*. Episode 6. Event occurs at 50 (34 minutes). BBC.

[8] J. P. Joule (1884), *The Scientific Paper of James Prescott Joule*, The Physical Society of London, p. 274, I am inclined to believe that both of these hypotheses will be found to hold good,—that in some instances, particularly in the case of sensible heat, or such as is indicated by the thermometer, heat will be found to consist in the living force of the particles of the bodies in which it is induced; whilst in others, particularly in the case of latent heat, the phenomena are produced by the separation of particle from particle, so as to cause them to attract one another through a greater space., Lecture on Matter, Living Force, and Heat. May 5 and 12, 1847

[9] Yaws' Handbook of Properties of the Chemical Elements 2011 Knovel

[10] Polynomial curve fits to Table 2.1. R. R. Rogers & M. K. Yau (1989). *A Short Course in Cloud Physics* (3rd ed.). Pergamon Press. p. 16. ISBN 0-7506-3215-1.

Chapter 6

Latent internal energy

The latent internal energy of a system is the internal energy a system requires to undergo a phase transition. Its value is specific to the substance or mix of substances in question. The value can also vary with temperature and pressure. Generally speaking the value is different for the type of phase change being accomplished. Examples can include Latent internal energy of vaporization (liquid to vapor), Latent internal energy of crystallization (liquid to solid) Latent internal energy of sublimation (solid to vapor). These values are usually expressed in units of energy per mole or per mass such as J/mol or BTU/lb. Often a negative sign will be used to represent energy being withdrawn from the system, while a positive value represents energy being added to the system. However, reference sources do vary so check the source to be sure.

For every type of latent internal energy there is an opposite. For example, the latent internal energy of fusion]] (liquid to solid) is equal to the negative of the Latent internal energy of melting (solid to liquid)

Chapter 7

Trouton's ratio

7.1 Physics

In physics, **Trouton's ratio** states that latent heat is connected to boiling point roughly by:

$$\frac{L_{vap}}{T_{boiling}} \approx 87 - 88 \, \frac{J}{Kmol}$$

7.2 Rheology

In rheology, **Trouton's ratio** is the ratio of extensional viscosity to shear viscosity.[1] For a Newtonian fluid Trouton's ratio is 3.[2]

7.3 See also

- Frederick Thomas Trouton
- Trouton's rule about entropy of vaporization

7.4 References

[1] http://web.mst.edu/~{}wlf/Mechanical/Trouton.html

[2] http://web.mit.edu/nnf/research/ere/ere.html

Chapter 8

Volatility (chemistry)

In chemistry and physics, **volatility** is the tendency of a substance to vaporize. Volatility is directly related to a substance's vapor pressure. At a given temperature, a substance with higher vapor pressure vaporizes more readily than a substance with a lower vapor pressure.[1][2][3][4]

The term is primarily written to be applied to liquids; however, it may be used to describe the process of sublimation which is associated with solid substances, such as dry ice (solid carbon dioxide) and ammonium chloride, which can change directly from the solid state to a vapor without becoming liquid.

8.1 Relations between vapor pressure, temperature, and boiling point

Main article: Vapor pressure

The vapor pressure of a substance is the pressure at which its gas phase is in equilibrium with its condensed phases (liquid or solid). It is a measure of the tendency of molecules and atoms to escape from a liquid or a solid. A liquid's atmospheric pressure boiling point corresponds to the temperature at which its vapor pressure is equal to the surrounding atmospheric pressure and it is often called the *normal boiling point*.

The higher the vapor pressure of a liquid at a given temperature, the higher the volatility and the lower the normal boiling point of the liquid. The vapor pressure chart (right hand side) displays the vapor pressures dependency for a variety of liquids as a function of temperature.[5]

For example, at any given temperature, methyl chloride has the highest vapor pressure of any of the liquids in the chart. It also has the lowest normal boiling point (−24.2 °C), which is where the vapor pressure curve of methyl chloride (the blue line) intersects the horizontal pressure line of one atmosphere (atm) of absolute vapor pressure.

8.2 See also

- Clausius–Clapeyron relation
- Distillation
- Fractional distillation
- Partial pressure
- Raoult's law
- Relative volatility

- Vapor–liquid equilibrium

- Volatile organic compound

8.3   References

[1] Gases and Vapor (University of Kentucky website)

[2] Definition of Terms (University of Victoria website)

[3] James G. Speight (2006). *The Chemistry and Technology of Petroleum* (4th ed.). CRC Press. ISBN 978-0-8493-9067-8.

[4] Kister, Henry Z. (1992-02-01). *Distillation Design* (1st ed.). McGraw-hill. ISBN 978-0-07-034909-4.

[5] Perry, R.H. and Green, D.W. (Editors); Don W. Green; James O. Maloney (1997). *Perry's Chemical Engineers' Handbook* (7th ed.). McGraw-Hill. ISBN 978-0-07-049841-9.

8.4   External links

- Volatility from ilpi.com

- Definition of volatile from Wiktionary

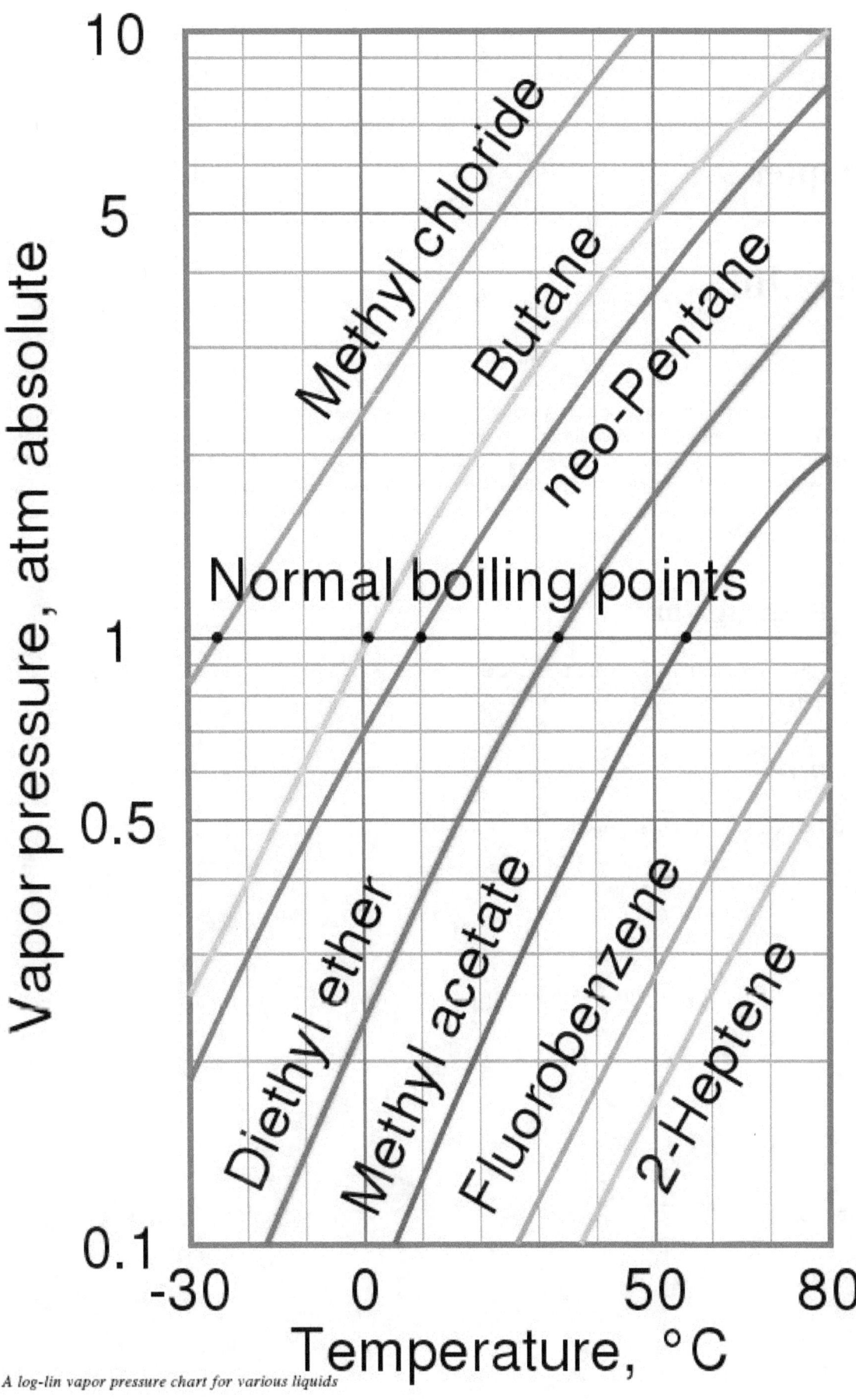

A log-lin vapor pressure chart for various liquids

Chapter 9

Binodal

In thermodynamics, the **binodal**, also known as the **coexistence curve** or **binodal curve**, denotes the condition at which two distinct phases may coexist. Equivalently, it is the boundary between the set of conditions in which it is thermodynamically favorable for the system to be fully mixed and the set of conditions in which it is thermodynamically favorable for it to phase separate.[1] In general, the binodal is defined by the condition at which the chemical potential of all solution components is equal in each phase. The extremum of a binodal curve in temperature coincides with the one of the spinodal curve and is known as a critical point.

9.1 Binary systems

In binary (two component) mixtures, the binodal can be determined at a given temperature by drawing a tangent line to the free energy.[1]

9.2 References

[1] IUPAC binodal curve definition http://old.iupac.org/goldbook/BT07273.pdf accessed 2/20/13

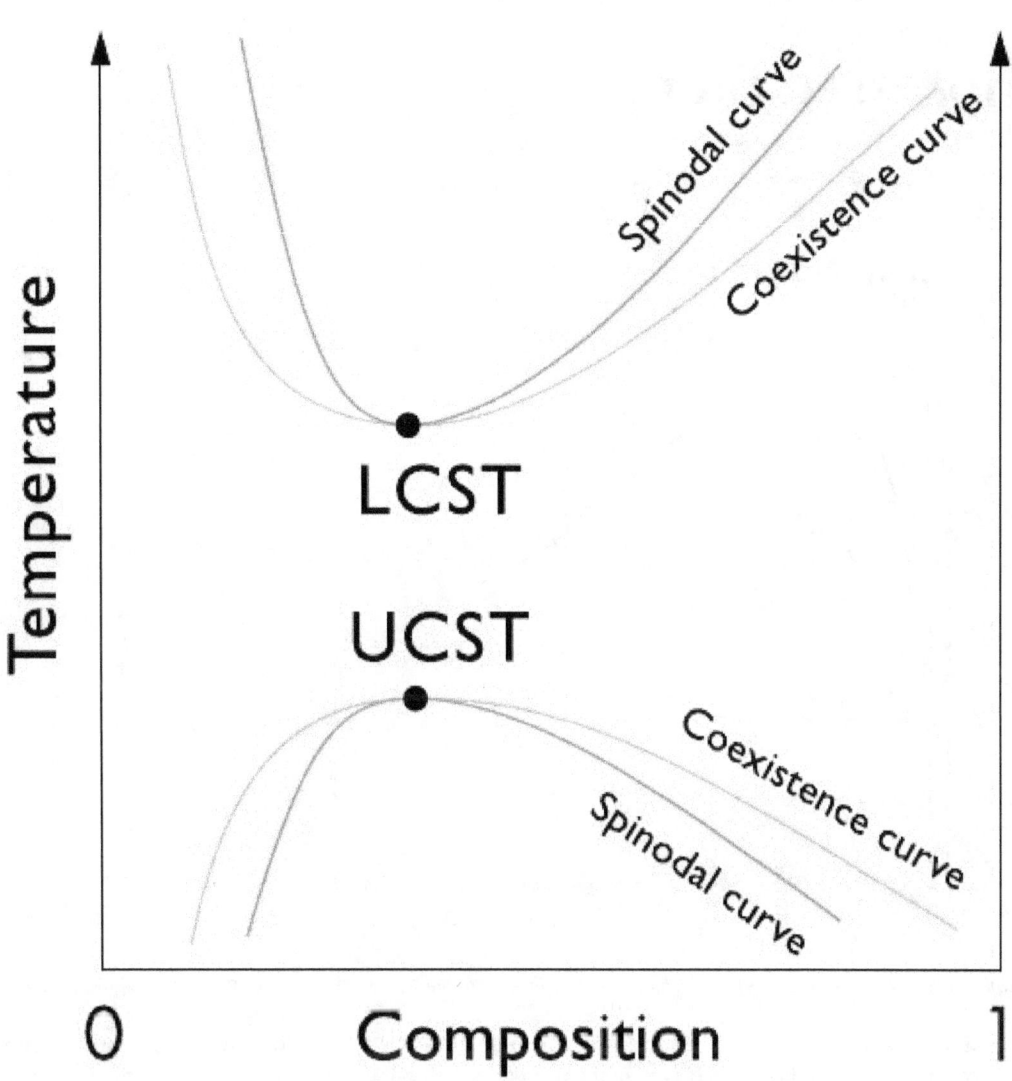

A phase diagram displaying binodal curves.

Chapter 10

Compressed fluid

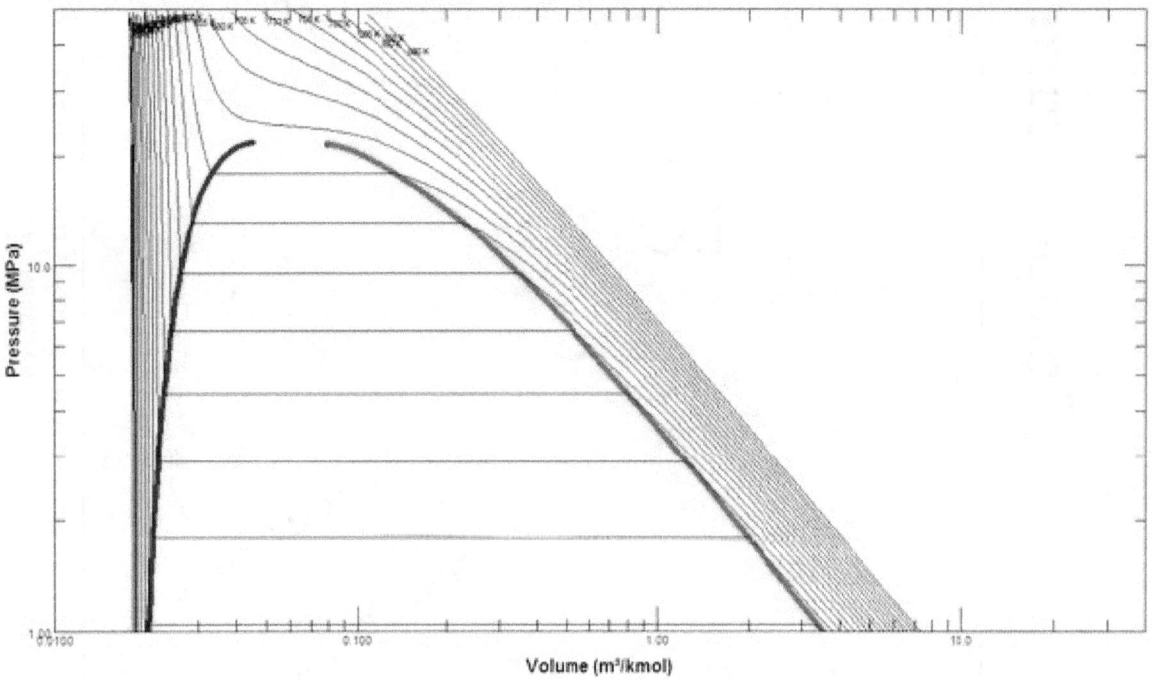

A P-v diagram for liquid water. The compressed fluid region is located to the left of the blue line (the liquid-vapor phase boundary).

A **compressed fluid** (also called a **subcooled fluid** or **subcooled liquid**) is a fluid under mechanical and or thermodynamic conditions that force it to be a liquid.[1] It is a liquid at a temperature lower than the saturation temperature at a given pressure. In a plot that compares absolute pressure and specific volume (commonly called a P-v diagram), of a real gas, a compressed fluid is to the left of the liquid-vapor phase boundary; that is, it will be to the left of the vapor dome.

Conditions that cause a fluid to be compressed include:

- Specific volume less than the specific volume of a saturated liquid

- Fluid temperature below the saturation temperature

- Pressure above the saturation pressure

- Enthalpy smaller than the enthalpy of a saturated liquid

The term *compressed liquid* emphasizes that the pressure is greater than the saturation pressure for the given temperature. Compressed liquid properties are relatively independent of pressure. As such, it is usually acceptable to treat a compressed liquid as a saturated liquid at the given temperature.

10.1 References

[1] "Thermodynamics: An Engineering Approach" by Yunus A. Çengel, Michael A. Boles, p.65, ISBN 0-07-121688-X

10.2 See also

- Gas cylinder

Chapter 11

Cooling curve

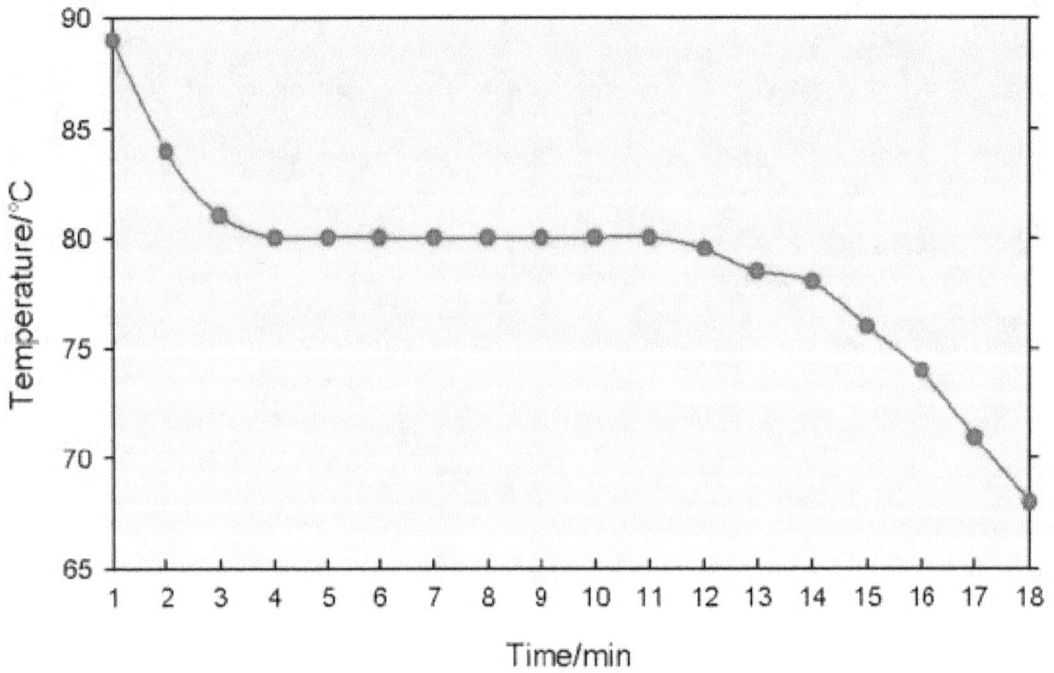

A cooling curve of naphthalene from liquid to solid.

A **cooling curve** is a line graph that represents the change of phase of matter, typically from a gas to a solid or a liquid to a solid. The independent variable (X-axis) is time and the dependent variable (Y-axis) is temperature.[1] Below is an example of a cooling curve used in castings.

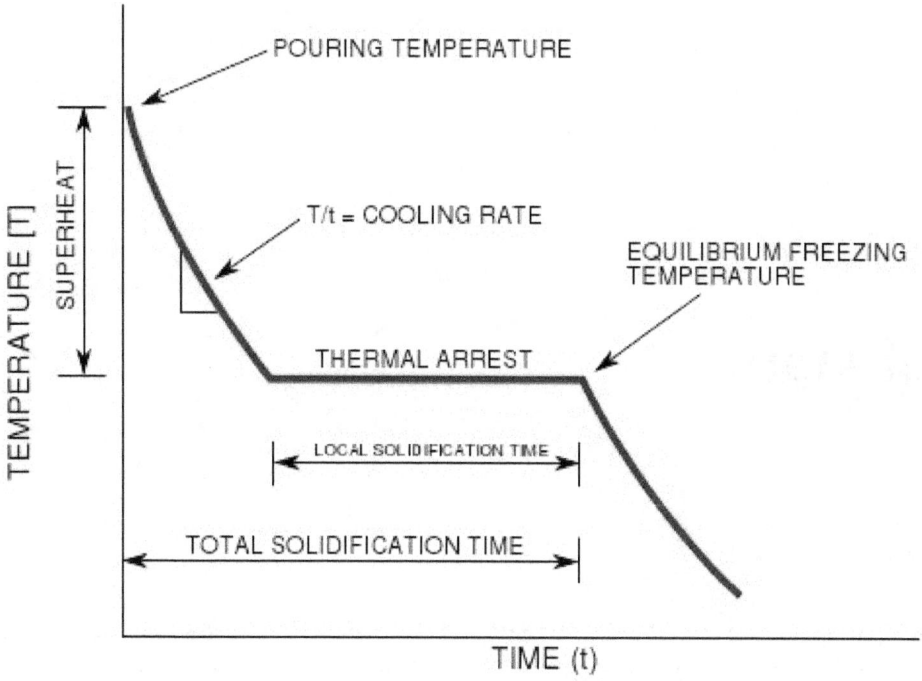

The initial point of the graph is the starting temperature of the matter, here noted as the "pouring temperature". When the phase change occurs there is a "thermal arrest", that is the temperature stays constant. This is because the matter has more internal energy as a liquid or gas than in the state that it is cooling to. The amount of energy required for a phase change is known as latent heat. The "cooling rate" is the slope of the cooling curve at any point.

In the part of the curve where the temperature decreases, the kinetic energy also decreases while the potential energy stays the same. However, at the phase transition, where the curve is flat, the kinetic energy stays the same while the potential energy decreases.

11.1 References

[1] Garland, Nibler, and Shoemaker. Experiments in Physical Chemistry (7th ed.)

Chapter 12

Equation of state

For the use of this concept in cosmology, see Equation of state (cosmology). For the use of this concept in optimal control theory, see Optimal control#General method.

In physics and thermodynamics, an **equation of state** is a relation between state variables.[1] More specifically, an equation of state is a thermodynamic equation describing the state of matter under a given set of physical conditions. It is a constitutive equation which provides a mathematical relationship between two or more state functions associated with the matter, such as its temperature, pressure, volume, or internal energy. Equations of state are useful in describing the properties of fluids, mixtures of fluids, solids, and even the interior of stars.

12.1 Overview

The most prominent use of an equation of state is to correlate densities of gases and liquids to temperatures and pressures. One of the simplest equations of state for this purpose is the ideal gas law, which is roughly accurate for weakly polar gases at low pressures and moderate temperatures. However, this equation becomes increasingly inaccurate at higher pressures and lower temperatures, and fails to predict condensation from a gas to a liquid. Therefore, a number of more accurate equations of state have been developed for gases and liquids. At present, there is no single equation of state that accurately predicts the properties of all substances under all conditions.

In addition, there are also equations of state describing solids, including the transition of solids from one crystalline state to another. There are equations that model the interior of stars, including neutron stars, dense matter (quark–gluon plasmas) and radiation fields. A related concept is the perfect fluid equation of state used in cosmology.

In practical context, the equations of state are instrumental for PVT calculation in process engineering problems and especially in petroleum gas/liquid equilibrium calculations. A successful PVT model based on a fitting equation of state can be helpful to determine the state of the flow regime, the parameters for handling the reservoir fluids, piping and sizing.

12.2 Historical

12.2.1 Boyle's law (1662)

Boyle's Law was perhaps the first expression of an equation of state. In 1662, the noted Irish physicist and chemist Robert Boyle performed a series of experiments employing a J-shaped glass tube, which was sealed on one end. Mercury was added to the tube, trapping a fixed quantity of air in the short, sealed end of the tube. Then the volume of gas was carefully measured as additional mercury was added to the tube. The pressure of the gas could be determined by the difference between the mercury level in the short end of the tube and that in the long, open end. Through these experiments, Boyle noted that the gas volume varied inversely with the pressure. In mathematical form, this can be stated as:

pV = constant.

The above relationship has also been attributed to Edme Mariotte and is sometimes referred to as **Mariotte's law**. However, Mariotte's work was not published until 1676.

12.2.2 Charles's law or Law of Charles and Gay-Lussac (1787)

In 1787 the French physicist Jacques Charles found that oxygen, nitrogen, hydrogen, carbon dioxide, and air expand to the same extent over the same 80 kelvin interval. Later, in 1802, Joseph Louis Gay-Lussac published results of similar experiments, indicating a linear relationship between volume and temperature:

$$\frac{V_1}{T_1} = \frac{V_2}{T_2}.$$

12.2.3 Dalton's law of partial pressures (1801)

Dalton's Law of partial pressure states that the pressure of a mixture of gases is equal to the sum of the pressures of all of the constituent gases alone.

Mathematically, this can be represented for n species as:

$$p_{total} = p_1 + p_2 + \cdots + p_n = p_{total} = \sum_{i=1}^{n} p_i.$$

12.2.4 The ideal gas law (1834)

In 1834 Émile Clapeyron combined Boyle's Law and Charles' law into the first statement of the *ideal gas law*. Initially the law was formulated as $pVm = R(TC + 267)$ (with temperature expressed in degrees Celsius), where R is the gas constant. However, later work revealed that the number should actually be closer to 273.2, and then the Celsius scale was defined with 0 °C = 273.15 K, giving:

$$pV_m = R(T_C + 273.15).$$

12.2.5 Van der Waals equation of state (1873)

In 1873, J. D. van der Waals introduced the first equation of state derived by the assumption of a finite volume occupied by the constituent molecules.[2] His new formula revolutionized the study of equations of state, and was most famously continued via the Redlich–Kwong equation of state and the Soave modification of Redlich-Kwong.

12.3 Major equations of state

For a given amount of substance contained in a system, the temperature, volume, and pressure are not independent quantities; they are connected by a relationship of the general form:

$$f(p, V, T) = 0.$$

In the following equations the variables are defined as follows. Any consistent set of units may be used, although SI units are preferred. Absolute temperature refers to use of the Kelvin (K) or Rankine (°R) temperature scales, with zero being absolute zero.

p = pressure (absolute)

V = volume

n = number of moles of a substance

$V_m = \frac{V}{n}$ = molar volume, the volume of 1 mole of gas or liquid

T = absolute temperature

R = ideal gas constant (8.3144621 J/(mol·K))

p_c = pressure at the critical point

V_c = molar volume at the critical point

T_c = absolute temperature at the critical point

12.3.1 Classical ideal gas law

The classical ideal gas law may be written:

$$pV = nRT.$$

In the form shown above, the equation of state is thus

$$f(p, V, T) = pV - nRT = 0 .$$

The ideal gas law may also be expressed as follows

$$p = \rho(\gamma - 1)e$$

where ρ is the density, $\gamma = C_p/C_v$ is the adiabatic index (ratio of specific heats), $e = C_v T$ is the internal energy per unit mass (the "specific internal energy"), C_v is the specific heat at constant volume, and C_p is the specific heat at constant pressure.

12.4 Cubic equations of state

Cubic equations of state are called such because they can be rewritten as a cubic function of V_m.

12.4.1 Van der Waals equation of state

The Van der Waals equation of state may be written:

$$\left(p + \frac{a}{V_m^2}\right)(V_m - b) = RT$$

where V_m is molar volume. The substance-specific constants a and b can be calculated from the critical properties p_c, T_c and V_c (noting that V_c is the molar volume at the critical point) as:

$$a = 3p_c V_c^2$$

$$b = \frac{V_c}{3}.$$

Also written as

$$a = \frac{27(RT_c)^2}{64p_c}$$

$$b = \frac{RT_c}{8p_c}.$$

Proposed in 1873, the van der Waals equation of state was one of the first to perform markedly better than the ideal gas law. In this landmark equation a is called the attraction parameter and b the repulsion parameter or the effective molecular volume. While the equation is definitely superior to the ideal gas law and does predict the formation of a liquid phase, the agreement with experimental data is limited for conditions where the liquid forms. While the van der Waals equation is commonly referenced in text-books and papers for historical reasons, it is now obsolete. Other modern equations of only slightly greater complexity are much more accurate.

The van der Waals equation may be considered as the ideal gas law, "improved" due to two independent reasons:

1. Molecules are thought as particles with volume, not material points. Thus V_m cannot be too little, less than some constant. So we get ($V_m - b$) instead of V_m .

2. While ideal gas molecules do not interact, we consider molecules attracting others within a distance of several molecules' radii. It makes no effect inside the material, but surface molecules are attracted into the material from the surface. We see this as diminishing of pressure on the outer shell (which is used in the ideal gas law), so we write ($p+$ something) instead of p . To evaluate this 'something', let's examine an additional force acting on an element of gas surface. While the force acting on each surface molecule is $\sim \rho$, the force acting on the whole element is $\sim \rho^2 \sim \frac{1}{V_m^2}$.

With the reduced state variables, i.e. $V_r = V_m/V_c$, $P_r = P/P_c$ and $T_r = T/T_c$, the reduced form of the Van der Waals equation can be formulated:

$$\left(P_r + \frac{3}{V_r^2} \right)(3V_r - 1) = 8T_r$$

The benefit of this form is that for given T_r and P_r, the reduced volume of the liquid and gas can be calculated directly using Cardano's method for the reduced cubic form:

$$V_r^3 - \left(\frac{1}{3} + \frac{8T_r}{3P_r} \right) V_r^2 + \frac{3V_r}{P_r} - \frac{1}{P_r} = 0$$

For $P_r<1$ and $T_r<1$, the system is in a state of vapor–liquid equilibrium. The reduced cubic equation of state yields in that case 3 solutions. The largest and the lowest solution are the gas and liquid reduced volume.

12.4.2   Redlich-Kwong equation of state

$$p = \frac{RT}{V_m - b} - \frac{a}{\sqrt{T}\,V_m\,(V_m + b)}$$

$$a = \frac{0.42748\,R^2\,T_c^{\,5/2}}{p_c}$$

$$b = \frac{0.08664\,RT_c}{p_c}$$

Introduced in 1949, the Redlich-Kwong equation of state was a considerable improvement over other equations of the time. It is still of interest primarily due to its relatively simple form. While superior to the van der Waals equation of state, it performs poorly with respect to the liquid phase and thus cannot be used for accurately calculating vapor–liquid equilibria. However, it can be used in conjunction with separate liquid-phase correlations for this purpose.

The Redlich-Kwong equation is adequate for calculation of gas phase properties when the ratio of the pressure to the critical pressure (reduced pressure) is less than about one-half of the ratio of the temperature to the critical temperature (reduced temperature):

$$\frac{p}{p_c} < \frac{T}{2T_c}.$$

12.4.3   Soave modification of Redlich-Kwong

$$p = \frac{RT}{V_m - b} - \frac{a\,\alpha}{V_m\,(V_m + b)}$$

$$a = \frac{0.427\,R^2\,T_c^2}{P_c}$$

$$b = \frac{0.08664\,RT_c}{P_c}$$

$$\alpha = \left(1 + \left(0.48508 + 1.55171\,\omega - 0.15613\,\omega^2\right)\left(1 - T_r^{0.5}\right)\right)^2$$

$$T_r = \frac{T}{T_c}$$

Where ω is the acentric factor for the species.

This formulation for α is due to Graboski and Daubert. The original formulation from Soave is:

$$\alpha = \left(1 + \left(0.48 + 1.574\,\omega - 0.176\,\omega^2\right)\left(1 - T_r^{0.5}\right)\right)^2$$

for hydrogen:

$$\alpha = 1.202\exp\left(-0.30288\,T_r\right).$$

We can also write it in the polynomial form, with:

$$A = \frac{a\,\alpha\,P}{R^2\,T^2}$$

$$B = \frac{b\,P}{RT}$$

then we have:

$$0 = Z^3 - Z^2 + Z\left(A - B - B^2\right) - AB$$

where R is the universal gas constant and Z=PV/(RT) is the compressibility factor.

In 1972 G. Soave[3] replaced the $1/\sqrt(T)$ term of the Redlich-Kwong equation with a function $\alpha(T,\omega)$ involving the temperature and the acentric factor (the resulting equation is also known as the Soave-Redlich-Kwong equation). The α function was devised to fit the vapor pressure data of hydrocarbons and the equation does fairly well for these materials.

Note especially that this replacement changes the definition of a slightly, as the T_c is now to the second power.

12.4.4   Peng–Robinson equation of state

$$p = \frac{RT}{V_m - b} - \frac{a\,\alpha}{V_m^2 + 2bV_m - b^2}$$

$$a = \frac{0.457235\,R^2\,T_c^2}{p_c}$$

$$b = \frac{0.077796\,R\,T_c}{p_c}$$

$$\alpha = \left(1 + \kappa\left(1 - T_r^{0.5}\right)\right)^2$$

$$\kappa = 0.37464 + 1.54226\,\omega - 0.26692\,\omega^2$$

$$T_r = \frac{T}{T_c}$$

In polynomial form:

$$A = \frac{a\alpha p}{R^2\,T^2}$$

$$B = \frac{bp}{RT}$$

$$Z^3 - (1 - B)\,Z^2 + \left(A - 2B - 3B^2\right)Z - \left(AB - B^2 - B^3\right) = 0$$

where ω is the acentric factor of the species, R is the universal gas constant and Z=PV/(RT) is compressibility factor.

The Peng–Robinson equation was developed in 1976 at The University of Alberta in order to satisfy the following goals:[1]

1. The parameters should be expressible in terms of the critical properties and the acentric factor.

2. The model should provide reasonable accuracy near the critical point, particularly for calculations of the compressibility factor and liquid density.

3. The mixing rules should not employ more than a single binary interaction parameter, which should be independent of temperature pressure and composition.

4. The equation should be applicable to all calculations of all fluid properties in natural gas processes.

For the most part the Peng–Robinson equation exhibits performance similar to the Soave equation, although it is generally superior in predicting the liquid densities of many materials, especially nonpolar ones. The departure functions of the Peng–Robinson equation are given on a separate article.

12.4.5   Peng–Robinson-Stryjek-Vera equations of state

PRSV1

A modification to the attraction term in the Peng–Robinson equation of state published by Stryjek and Vera in 1986 (PRSV) significantly improved the model's accuracy by introducing an adjustable pure component parameter and by modifying the polynomial fit of the acentric factor.[5]

The modification is:

$$\kappa = \kappa_0 + \kappa_1 \left(1 + T_r^{0.5}\right)(0.7 - T_r)$$

$$\kappa_0 = 0.378893 + 1.4897153\,\omega - 0.17131848\,\omega^2 + 0.0196554\,\omega^3$$

where κ_1 is an adjustable pure component parameter. Stryjek and Vera published pure component parameters for many compounds of industrial interest in their original journal article. At reduced temperatures above 0.7, they recommend to set $\kappa_1 = 0$ and simply use $\kappa = \kappa_0$. For alcohols and water the value of κ_1 may be used up to the critical temperature and set to zero at higher temperatures.[5]

PRSV2

A subsequent modification published in 1986 (PRSV2) further improved the model's accuracy by introducing two additional pure component parameters to the previous attraction term modification.[6]

The modification is:

$$\kappa = \kappa_0 + \left[\kappa_1 + \kappa_2 \left(\kappa_3 - T_r\right)\left(1 - T_r^{0.5}\right)\right]\left(1 + T_r^{0.5}\right)(0.7 - T_r)$$

$$\kappa_0 = 0.378893 + 1.4897153\,\omega - 0.17131848\,\omega^2 + 0.0196554\,\omega^3$$

where κ_1, κ_2, and κ_3 are adjustable pure component parameters.

PRSV2 is particularly advantageous for VLE calculations. While PRSV1 does offer an advantage over the Peng–Robinson model for describing thermodynamic behavior, it is still not accurate enough, in general, for phase equilibrium calculations.[5] The highly non-linear behavior of phase-equilibrium calculation methods tends to amplify what would otherwise be acceptably small errors. It is therefore recommended that PRSV2 be used for equilibrium calculations when applying these models to a design. However, once the equilibrium state has been determined, the phase specific thermodynamic values at equilibrium may be determined by one of several simpler models with a reasonable degree of accuracy.[6]

One thing to note is that in the PRSV equation, the parameter fit is done in a particular temperature range which is usually below the critical temperature. Above the critical temperature, the PRSV alpha function tends to diverge and become arbitrarily large instead of tending towards 0. Because of this, alternate equations for alpha should be employed above the critical point. This is especially important for systems containing hydrogen which is often found at temperatures far above its critical point. Several alternate formulations have been proposed. Some well known ones are by Twu et all or by Mathias and Copeman.

12.4.6   Elliott, Suresh, Donohue equation of state

The Elliott, Suresh, and Donohue (ESD) equation of state was proposed in 1990.[7] The equation seeks to correct a shortcoming in the Peng–Robinson EOS in that there was an inaccuracy in the van der Waals repulsive term. The EOS accounts for the effect of the shape of a non-polar molecule and can be extended to polymers with the addition of an extra term (not shown). The EOS itself was developed through modeling computer simulations and should capture the essential physics of the size, shape, and hydrogen bonding.

$$\frac{pV_m}{RT} = Z = 1 + Z^{\text{rep}} + Z^{\text{att}}$$

where:

$$Z^{\text{rep}} = \frac{4c\eta}{1 - 1.9\eta}$$

$$Z^{\text{att}} = -\frac{z_m q\eta Y}{1 + k_1\eta Y}$$

and

c is a "shape factor", with $c = 1$ for spherical molecules

For non-spherical molecules, the following relation is suggested:

$c = 1 + 3.535\omega + 0.533\omega^2$ where ω is the acentric factor

The reduced number density η is defined as $\eta = \frac{v^* n}{V}$

where

v^*

n

V

The characteristic size parameter is related to the shape parameter c through

$$v^* = \frac{kT_c}{P_c}\Phi$$

where

$\Phi = \frac{0.0312 + 0.087(c-1) + 0.008(c-1)^2}{1.000 + 2.455(c-1) + 0.732(c-1)^2}$ and k is Boltzmann's constant.

Noting the relationships between Boltzmann's constant and the Universal gas constant, and observing that the number of molecules can be expressed in terms of Avogadro's number and the molar mass, the reduced number density η can be expressed in terms of the molar volume as

$$\eta = \frac{RT_c}{P_c}\Phi\frac{1}{V_m}.$$

The shape parameter q appearing in the Attraction term and the term Y are given by

$$q = 1 + k_3(c - 1)$$

$$Y = \exp\left(\frac{\epsilon}{kT}\right) - k_2$$

where ϵ is the depth of the square-well potential and is given by

$$\frac{\epsilon}{k} = \frac{1.000 + 0.945(c - 1) + 0.134(c - 1)^2}{1.023 + 2.225(c - 1) + 0.478(c - 1)^2}$$

z_m , k_1 , k_2 and k_3 are constants in the equation of state:

$z_m = 9.49$ for spherical molecules (c=1)

$k_1 = 1.7745$ for spherical molecules (c=1)

$k_2 = 1.0617$ for spherical molecules (c=1)

$k_3 = 1.90476.$

The model can be extended to associating components and mixtures of nonassociating components. Details are in the paper by J.R. Elliott, Jr. *et al.* (1990).[7]

12.5 Non-cubic equations of state

12.5.1 Dieterici equation of state

$$p(V - b) = RTe^{-a/RTV}$$

where a is associated with the interaction between molecules and b takes into account the finite size of the molecules, similar to the Van der Waals equation.

The reduced coordinates are:

$$T_c = \frac{a}{4Rb}, \quad p_c = \frac{a}{4b^2e^2}, \quad V_c = 2b.$$

12.6 Virial equations of state

12.6.1 Virial equation of state

Main article: Virial expansion

$$\frac{pV_m}{RT} = 1 + \frac{B}{V_m} + \frac{C}{V_m^2} + \frac{D}{V_m^3} + \cdots$$

$$B = -V_c$$

$$C = \frac{V_c^2}{9}$$

Although usually not the most convenient equation of state, the virial equation is important because it can be derived directly from statistical mechanics. This equation is also called the Kamerlingh Onnes equation. If appropriate assumptions are made about the mathematical form of intermolecular forces, theoretical expressions can be developed for each of the coefficients. In this case B corresponds to interactions between pairs of molecules, C to triplets, and so on. Accuracy can be increased indefinitely by considering higher order terms. The coefficients B, C, D, etc. are functions of temperature only.

It can also be used to work out the Boyle Temperature (the temperature at which B = 0 and ideal gas laws apply) from a and b from the Van der Waals equation of state, if you use the value for B shown below:

$$B = b - \frac{a}{RT}.$$

12.6.2 The BWR equation of state

Main article: Benedict–Webb–Rubin equation

$$p = \rho RT + \left(B_0 RT - A_0 - \frac{C_0}{T^2} + \frac{D_0}{T^3} - \frac{E_0}{T^4} \right) \rho^2 + \left(bRT - a - \frac{d}{T} \right) \rho^3 + \alpha \left(a + \frac{d}{T} \right) \rho^6 + \frac{c\rho^3}{T^2} \left(1 + \gamma\rho^2 \right) \exp\left(-\gamma\rho^2 \right)$$

where

p = pressure

ϱ = the molar density

Values of the various parameters for 15 substances can be found in K.E. Starling (1973). *Fluid Properties for Light Petroleum Systems*. Gulf Publishing Company.

12.7 Multiparameter equations of state

12.7.1 Helmholtz Function form

Multiparameter equations of state (MEOS) can be used to represent pure fluids with high accuracy, in both the liquid and gaseous states. MEOS's represent the Helmholtz function of the fluid as the sum of ideal gas and residual terms. Both terms are explicit in reduced temperature and reduced density - thus:

$$\frac{a(T,\rho)}{RT} = \frac{a^o(T,\rho) + a^r(T,\rho)}{RT} = \alpha^o(\tau,\delta) + \alpha^r(\tau,\delta)$$

Where:

$$\tau = \frac{T_r}{T}, \delta = \frac{\rho}{\rho_r}$$

The reduced density and temperature are typically, though not always, the critical values for the pure fluid. Other thermodynamic functions can be derived from the MEOS by using appropriate derivatives of the Helmholtz function; hence, because integration of the MEOS is not required, there are few restrictions as to the functional form of the ideal or residual terms. Typical MEOS use upwards of 50 fluid specific parameters, but are able to represent the fluid's properties with high accuracy. MEOS are available currently for about 50 of the most common industrial fluids including refrigerants. Mixture models also exist.

12.8 Other equations of state of interest

12.8.1 Stiffened equation of state

When considering water under very high pressures (typical applications are underwater nuclear explosions, sonic shock lithotripsy, and sonoluminescence) the stiffened equation of state is often used:

$$p = \rho(\gamma - 1)e - \gamma p^0$$

where e is the internal energy per unit mass, γ is an empirically determined constant typically taken to be about 6.1, and p^0 is another constant, representing the molecular attraction between water molecules. The magnitude of the correction is about 2 gigapascals (20,000 atmospheres).

The equation is stated in this form because the speed of sound in water is given by $c^2 = \gamma(p + p^0)/\rho$.

Thus water behaves as though it is an ideal gas that is *already* under about 20,000 atmospheres (2 GPa) pressure, and explains why water is commonly assumed to be incompressible: when the external pressure changes from 1 atmosphere to 2 atmospheres (100 kPa to 200 kPa), the water behaves as an ideal gas would when changing from 20,001 to 20,002 atmospheres (2000.1 MPa to 2000.2 MPa).

This equation mispredicts the specific heat capacity of water but few simple alternatives are available for severely non-isentropic processes such as strong shocks.

12.8.2   Ultrarelativistic equation of state

An ultrarelativistic fluid has equation of state

$$p = \rho_m c_s^2$$

where p is the pressure, ρ_m is the mass density, and c_s is the speed of sound.

12.8.3   Ideal Bose equation of state

The equation of state for an ideal Bose gas is

$$pV_m = RT \frac{\mathrm{Li}_{\alpha+1}(z)}{\zeta(\alpha)} \left(\frac{T}{T_c} \right)^{\alpha}$$

where α is an exponent specific to the system (e.g. in the absence of a potential field, $\alpha=3/2$), z is $\exp(\mu/kT)$ where μ is the chemical potential, Li is the polylogarithm, ζ is the Riemann zeta function, and Tc is the critical temperature at which a Bose–Einstein condensate begins to form.

12.8.4   Jones-Wilkins-Lee equation of state for explosives (JWL-equation)

The equation of state from Jones-Wilkins-Lee is used to describe the detonation products of explosives.

$$p = A \cdot \left(1 - \frac{\omega}{R_1 \cdot V} \right) \cdot \exp(-R_1 \cdot V) + B \cdot \left(1 - \frac{\omega}{R_2 \cdot V} \right) \cdot \exp(-R_2 \cdot V) + \frac{\omega \cdot e_0}{V}$$

The ratio $V = \rho_e/\rho$ is defined by using ρ_e = density of the explosive (solid part) and ρ = density of the detonation products. The parameters A, B, R_1, R_2 and ω are given by several references.[8] In addition, the initial density (solid part) ρ_0, speed of detonation V_D, Chapman–Jouguet pressure P_{CJ} and the chemical energy of the explosive e_0 are given in such references. These parameters are obtained by fitting the JWL-EOS to experimental results. Typical parameters for some explosives are listed in the table below.

12.9   Equations of state for solids and liquids

Common abbreviations: $\eta := (V/V_0)^{1/3}$, $K_0' := \frac{dK_0}{dp}$

- Tait equation for water and other liquids. Several equations are referred to as the **Tait equation**.

- Murnaghan equation of state

$$p(V) = \frac{K_0}{K_0'} \left[\eta^{-3K_0'} - 1 \right]$$

- Birch–Murnaghan equation of state

$$p(V) = \frac{3K_0}{2}\left(\frac{1-\eta^2}{\eta^7}\right)\left\{1 + \frac{3}{4}\left(K_0' - 4\right)\left(\frac{1-\eta^2}{\eta^2}\right)\right\}$$

- **Stacey-Brennan-Irvine equation of state**[10] (falsely often refer to Rose-Vinet equation of state

$$p(V) = 3K_0\left(\frac{1-\eta}{\eta^2}\right)\exp\left[\frac{3}{2}\left(K_0' - 1\right)\left(1-\eta\right)\right]$$

- **Modified Rydberg equation of state**[11][12][13] (more reasonable form for strong compression)

$$p(V) = 3K_0\left(\frac{1-\eta}{\eta^5}\right)\exp\left[\frac{3}{2}\left(K_0' - 3\right)\left(1-\eta\right)\right]$$

- **Adapted Polynomial equation of state**[14] (second order form = AP2, adapted for extreme compression)

$$p(V) = 3K_0\left(\frac{1-\eta}{\eta^5}\right)\exp\left[c_0(1-\eta)\right]\left\{1 + c_2\eta(1-\eta)\right\}$$

with

$$c_0 = -\ln\left(\frac{3K_0}{p_{FG0}}\right)\ ,\quad p_{FG0} = a_0\left(\frac{Z}{V_0}\right)^{\frac{5}{3}},\quad c_2 = \frac{3}{2}\left(K_0' - 3\right) - c_0$$

where a_0 = 0.02337 GPa.nm^5. The total number of electrons Z in the initial volume V_0 determines the Fermi gas pressure p_{FG0} , which provides for the correct behavior at extreme compression. So far there are no known "simple" solids that require higher order terms.

- **Adapted polynomial equation of state**[14] (third order form = AP3)

$$p(V) = 3K_0\left(\frac{1-\eta}{\eta^5}\right)\exp\left[c_0(1-\eta)\right]\left\{1 + c_2\eta(1-\eta) + c_3\eta(1-\eta)^2\right\}$$

- Johnson–Holmquist equation of state

$$p(V) = \begin{cases} k_1\,\xi + k_2\,\xi^2 + k_3\,\xi^3 + \Delta p & \text{Compression} \\ k_1\,\xi & \text{Tension} \end{cases}\ ;\quad \xi := \frac{V_0}{V} - 1$$

- Mie–Gruneisen equation of state (for a more detailed discussion see ref.[15])

$$p(V) - p_0 = \frac{\Gamma}{V}(e - e_0)$$

- Anton-Schmidt equation of state

$$p(V) = -\beta\left(\frac{V}{V_0}\right)^n \ln\left(\frac{V}{V_0}\right)$$

where $\beta = K_0$ is the bulk modulus at equilibrium volume V_0 and $n = -\frac{K_0'}{2}$ typically about -2 is often related to the Grüneisen parameter by $n = -\frac{1}{6} - \gamma_G$

12.10   See also

- Gas laws

- Departure function

- Table of thermodynamic equations

- Real gas

- Cluster Expansion

12.11   References

[1] Perrot, Pierre (1998). *A to Z of Thermodynamics*. Oxford University Press. ISBN 0-19-856552-6.

[2] van der Waals, J. D. (1873). *On the Continuity of the Gaseous and Liquid States (doctoral dissertation)*. Universiteit Leiden.

[3] Soave, G. Equilibrium Constants from a Modified Redlich-Kwong Equation of State, Chem. Eng. Sci., 1 9 7 2, 27, 1197-1203

[4] Peng, D. Y., and Robinson, D. B. (1976). "A New Two-Constant Equation of State". *Industrial and Engineering Chemistry: Fundamentals* 15: 59–64. doi:10.1021/i160057a011.

[5] Stryjek, R. and Vera, J. H. (1986). "PRSV: An improved Peng–Robinson equation of state for pure compounds and mixtures". *The Canadian Journal of Chemical Engineering* 64: 323–333. doi:10.1002/cjce.5450640224.

[6] Stryjek, R. and Vera, J. H. (1986). "PRSV2: A cubic equation of state for accurate vapor—liquid equilibria calculations". *The Canadian Journal of Chemical Engineering* 64: 820–826. doi:10.1002/cjce.5450640516.

[7] J. Richard, Jr. Elliott, S. Jayaraman Suresh, Marc D. Donohue (1990). "A Simple Equation of State for Nonspherical and Associating Molecules". *Ind. Eng. Chem. Res.* 29 (7): 1476–1485. doi:10.1021/ie00103a057.

[8] B. M. Dobratz, P. C. Crawford (1985). *LLNL Explosives Handbook: Properties of Chemical Explosives and Explosive Simulants*. University of California; Lawrence Livermore National Laboratory; Report UCRL-5299; Rev.2;.

[9] Wilkins, Mark L. (1999), *Computer Simulation of Dynamic Phenomena*, Springer, p. 80

[10] Stacey, F. D.; Brennan, B. J.; Irvine, R. D. (1981). "Finite strain theories and comparisons with seismological data". *Surveys in Geophysics 4 (4): 189–232.*

[11] Holzapfel, W. B. (1991). *"Equations of states and scaling rules for molecular solids under strong compression" in "Molecular systems under high pressure" ed. R. Pucci and G. Piccino*. North-Holland: Elsevier. pp. 61–68.

[12] Holzapfel, W. B. (1991). "Equations of state for solids under strong compression". *High Press. Res.* 7, 290-293.

[13] Holzapfel, Wi. B. (1996). "Physics of solids under strong compression". *Rep. Prog. Phys.* 59, 29-90.

[14] Holzapfel, W. B. (1998). "Equation of state for solids under strong compression". *High Press. Res. 16, 81-126.*

[15] Holzapfel, W. B. (2004). *"Equations of state and thermophysical properties of solids under pressure" in "High-Pressure Crystallography" ed. A. Katrusiak and P. McMillan*. Netherlands: Kluver Academic. pp. 217–236.

- Elliot & Lira, (1999). *Introductory Chemical Engineering Thermodynamics*, Prentice Hall.

Chapter 13

Leidenfrost effect

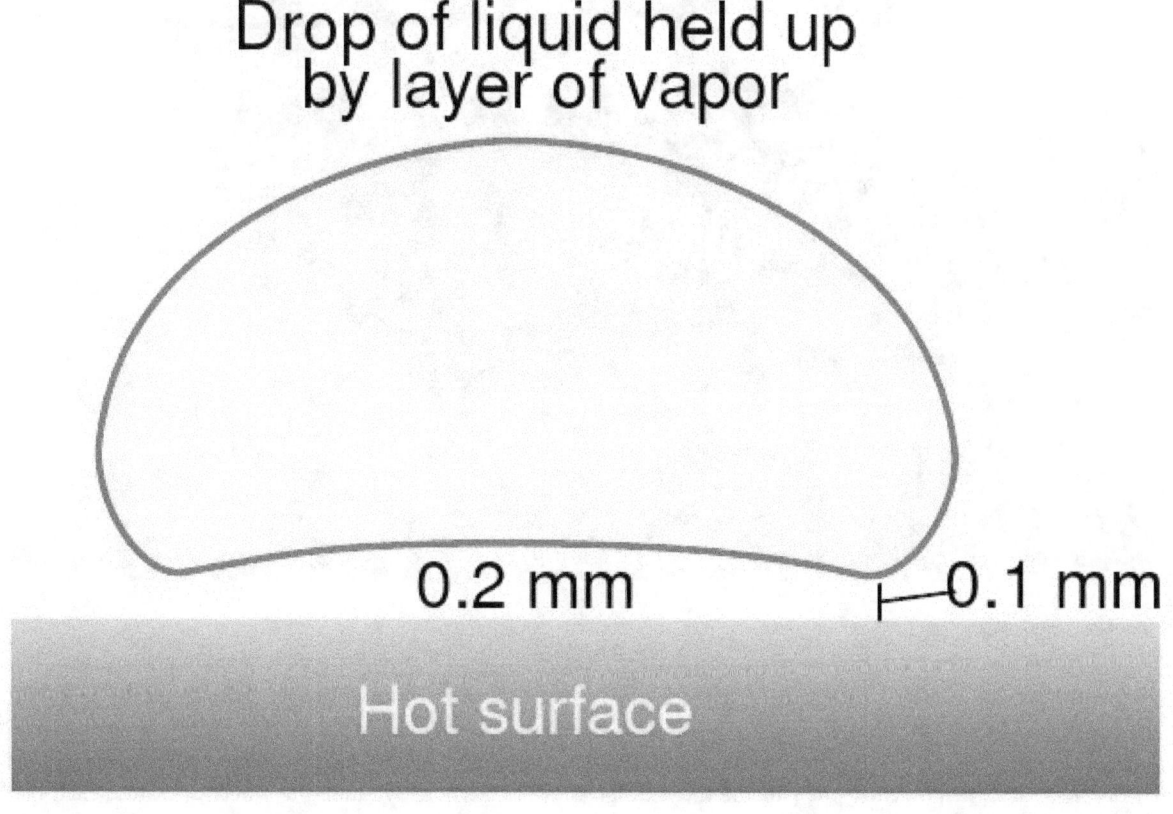

Drop of liquid held up
by layer of vapor

0.2 mm   ⊢0.1 mm

Hot surface

Leidenfrost droplet

The **Leidenfrost effect** is a physical phenomenon in which a liquid, in near contact with a mass significantly hotter than the liquid's boiling point, produces an insulating vapor layer keeping that liquid from boiling rapidly. Due to this 'repulsive force,' the droplet hovers over the surface rather than making physical contact with it. This is most commonly seen when cooking; one sprinkles drops of water in a pan to gauge its temperature: if the pan's temperature is at or above the Leidenfrost point, the water skitters across the pan and takes longer to evaporate than in a pan below the temperature of the Leidenfrost point (but still above boiling temperature). The effect is also responsible for the ability of liquid nitrogen to skitter across floors. It has also been used in some potentially dangerous demonstrations, such as dipping a wet finger in molten lead[1] or blowing out a mouthful of liquid nitrogen, both enacted without injury to the demonstrator.[2] The latter is potentially lethal, particularly should one accidentally swallow the liquid nitrogen.[3]

It is named after Johann Gottlob Leidenfrost, who discussed it in *A Tract About Some Qualities of Common Water* in 1756.

13.1   Effect

A video clip demonstrating the Leidenfrost effect

The effect can be seen as drops of water are sprinkled onto a pan at various times as it heats up. Initially, as the temperature of the pan is just below 100 °C (212 °F), the water flattens out and slowly evaporates, or if the temperature of the pan is well below 100 °C (212 °F), the water stays liquid. As the temperature of the pan goes above 100 °C (212 °F), the water droplets hiss when touching the pan and these droplets evaporate quickly. Later, as the temperature exceeds the Leidenfrost point, the Leidenfrost effect comes into play. On contact with the pan, the water droplets bunch up into small balls of water and skitter around, lasting much longer than when the temperature of the pan was lower. This effect works until a much higher temperature causes any further drops of water to evaporate too quickly to cause this effect.

This is because at temperatures above the Leidenfrost point, the bottom part of the water droplet vaporizes immediately on contact with the hot plate. The resulting gas suspends the rest of the water droplet just above it, preventing any further direct contact between the liquid water and the hot plate. As steam has much poorer thermal conductivity, further heat transfer between the pan and the droplet is slowed down dramatically. This also results in the drop being able to skid around the pan on the layer of gas just under it.

The temperature at which the Leidenfrost effect begins to occur is not easy to predict. Even if the volume of the drop

Excitation of normal modes in a drop of water during the Leidenfrost effect

of liquid stays the same, the Leidenfrost point may be quite different, with a complicated dependence on the properties of the surface, as well as any impurities in the liquid. Some research has been conducted into a theoretical model of the system, but it is quite complicated.[4] As a very rough estimate, the Leidenfrost point for a drop of water on a frying pan might occur at 193 °C (379 °F).

The effect was also described by the eminent Victorian steam boiler designer, Sir William Fairbairn, in reference to its effect on massively reducing heat transfer from a hot iron surface to water, such as within a boiler. In a pair of lectures on boiler design,[5] he cited the work of Pierre Hippolyte Boutigny (1798-1884) and Professor Bowman of King's College, London in studying this. A drop of water that was vaporized almost immediately at 168 °C (334 °F) persisted for 152 seconds at 202 °C (396 °F). Lower temperatures in a boiler firebox might evaporate water more quickly as a result; compare Mpemba effect. An alternative approach was to increase the temperature beyond the Leidenfrost point. Fairbairn considered this too, and may have been contemplating the flash steam boiler, but considered the technical aspects insurmountable for the time.

The Leidenfrost point may also be taken to be the temperature for which the hovering droplet lasts longest.[6]

It has been demonstrated that it is possible to stabilize the Leidenfrost vapour layer of water by exploiting superhydrophobic surfaces. In this case, once the vapour layer is established, cooling never collapses the layer, and no nucleate boiling occurs; the layer instead slowly relaxes until the surface is cooled.[7]

Leidenfrost effect has been used for the development of high sensitivity ambient mass spectrometry. Under the influence of Leidenfrost condition the Levitating droplet does not release molecules out and the molecules are enriched inside the droplet. At the last moment of droplet evaporation all of the enriched molecules release in a short time domain and thus increase the sensitivity.[8]

Heat transfer for water (@ 1 atm)

S-shaped graph when heat flux (q") is compared to temperature.

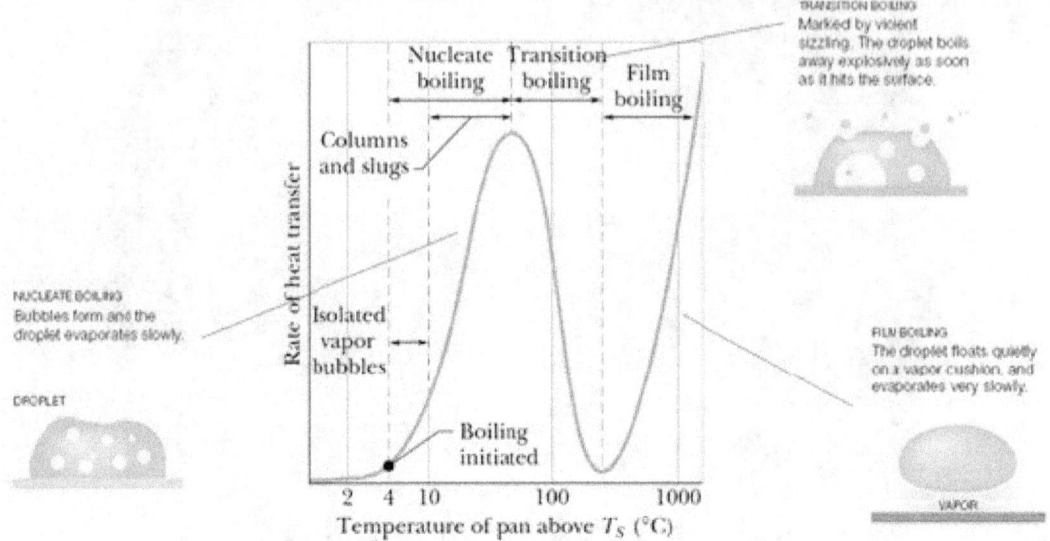

Behavior of water on a hot plate. Graph shows heat transfer (flux) vs temperature. Leidenfrost effect occurs after transition boiling.

A heat engine based on the Leidenfrost effect has been prototyped. It has the advantage of extremely low friction.[9]

13.2   Leidenfrost point

The Leidenfrost point signifies the onset of stable film boiling. It represents the point on the boiling curve where the heat flux is at the minimum and the surface is completely covered by a vapor blanket. Heat transfer from the surface to the liquid occurs by conduction and radiation through the vapor. In 1756, Leidenfrost observed that water droplets supported by the vapor film slowly evaporate as they move about on the hot surface. As the surface temperature is increased, radiation through the vapor film becomes more significant and the heat flux increases with increasing excess temperature.

The minimum heat flux for a large horizontal plate can be derived from Zuber's equation,[6]

$$\frac{q}{A}_{min} = Ch_{fg}\rho_v \left[\frac{\sigma g(\rho_L - \rho_v)}{(\rho_L + \rho_v)^2} \right]^{1/4}$$

where the properties are evaluated at saturation temperature. Zuber's constant, C is approximately 0.09 for most fluids at moderate pressures.

13.3   Heat transfer correlations

The heat transfer coefficient may be approximated using Bromley's equation,[6]

$$h = C \left[\frac{k_v^3 \rho_v g(\rho_L - \rho_v)(h_{fg} + 0.4c_{pv}(T_s - T_{sat}))}{D_o \mu_v (T_s - T_{sat})} \right]^{1/4}$$

Where, D_o is the outside diameter of the tube. The correlation constant C is 0.62 for horizontal cylinders and vertical plates and 0.67 for spheres. Vapor properties are evaluated at film temperature.

A water droplet experiencing Leidenfrost effect on a hot stove plate

For stable film boiling on a horizontal surface, Berenson has modified Bromley's equation to yield,[10]

$$h = 0.325 \left[\frac{k_{vf}^3 \rho_{vf} g(\rho_L - \rho_v)(h_{fg} + 0.4 c_{pv}(T_s - T_{sat}))}{\mu_{vf}(T_s - T_{sat})\sqrt{\sigma/g(\rho_L - \rho_v)}} \right]^{1/4}$$

For vertical tubes, Hsu and Westwater have correlated the following equation,[10]

$$h \left[\frac{\mu_v^2}{g\rho_v(\rho_L - \rho_v)k_v^3} \right]^{1/3} = 0.0020 \left[\frac{4m}{\pi D_v \mu_v} \right]^{0.6}$$

Where, m is the mass flow rate in lb_m/hr at the upper end of the tube

At excess temperatures above that at the minimum heat flux, the contribution of radiation becomes appreciable and becomes dominant at high excess temperatures. The total heat transfer coefficient can be is thus a combination of the two. Bromley has suggested the following equations for film boiling boiling from the outer surface of horizontal tubes.

$$h^{4/3} = h_{conv}^{4/3} + h_{rad}h^{1/3}$$

If $h_{rad} < h_{conv}$,

$$h = h_{conv} + \frac{3}{4}h_{rad}$$

The effective radiation coefficient, h_{rad} can be expressed as,

$$h_{rad} = \frac{\varepsilon\sigma(T_s^4 - T_{sat}^4)}{(T_s - T_{sat})}$$

Where, ε is the emissivity of the solid and σ is the Stefan-Boltzmann constant.

13.4   The pressure field in a Leidenfrost droplet

The equation for the pressure field in the vapor region between the droplet and the solid surface can be solved for using the standard momentum and continuity equations. For the sake of simplicity in solving, a linear temperature profile and a parabolic velocity profile are assumed within the vapor phase. The heat transfer within the vapor phase is assumed to be through conduction. With these approximations, the Navier-Stokes equation can be solved[11] to get the pressure field.

13.5   Leidenfrost temperature and surface tension effects

The Leidenfrost temperature is the property of a given set of solid-liquid pair. The temperature of the solid surface beyond which the liquid undergoes Leidenfrost phenomenon is termed as Leidenfrost temperature. The calculation of Leidenfrost temperature involves the calculation of minimum film boiling temperature of a fluid. Berenson[12] obtained a relation for the minimum film boiling temperature from minimum heat flux arguments. While the equation for the minimum film boiling temperature, which can be found in the reference above, is quite complex, the features of it can be understood from a physical perspective. One critical parameter to consider is the surface tension. The proportional relationship between the minimum film boiling temperature and surface tension is to be expected since fluids with higher surface tension need higher quantities of heat flux for the onset of nucleate boiling. Since film boiling occurs after nucleate boiling, the minimum temperature for film boiling should have a proportional dependence on the surface tension.

Henry[13] developed a model for Leidenfrost phenomenon which includes transient wetting and microlayer evaporation. Since the Leidenfrost phenomenon is a special case of film boiling, the Leidenfrost temperature is related to the minimum film boiling temperature via a relation which factors in the properties of the solid being used. While the Leidenfrost temperature is not directly related to the surface tension of the fluid, it is indirectly dependent on it through the film boiling temperature. For fluids with similar thermophysical properties, the one with higher surface tension usually has a higher Leidenfrost temperature.

For example, for saturated water-copper interface, the Leidenfrost temperature is 257 °C (495 °F). The Leidenfrost temperatures for glycerol and common alcohols are significantly smaller due to their lower surface tension values (density and viscosity differences are also contributing factors.)

13.6   Reactive Leidenfrost effect

Non-volatile materials were discovered in 2015 to also exhibit a 'reactive Leidenfrost effect,' whereby solid particles were observed to float above hot surfaces and skitter around erratically.[14] Detailed characterization of the reactive Leidenfrost effect was completed for small particles of cellulose (~0.5 mm) on high temperature polished surfaces by high speed photography. Cellulose was shown to decompose to short-chain oligomers which melt and wet smooth surfaces with increasing heat transfer associated with increasing surface temperature. Above 675 °C (1,247 °F), cellulose was observed to exhibit transition boiling with violent bubbling and associated reduction in heat transfer. Liftoff of the cellulose droplet (depicted at the right) was observed to occur above about 750 °C (1,380 °F) associated with a dramatic reduction in heat transfer.[15] High speed photography of the reactive Leidenfrost effect of cellulose on porous surfaces (macroporous alumina) was also shown to suppress the reactive Leidenfrost effect and enhance overall heat transfer rates to the particle from the surface. The new phenomenon of a 'reactive Leidenfrost (RL) effect' was characterized by a dimensionless quantity (φRL= τ_{conv}/τ_{rxn}), which relates the time constant of solid particle heat transfer to the time constant of particle reaction, with the reactive Leidenfrost effect occurring for $10^{-1} < \varphi$RL$ < 10^{+1}$. The reactive Leidenfrost effect with cellulose will occur in numerous high temperature applications with carbohydrate polymers including biomass conversion to biofuels, preparation and cooking of food, and tobacco use.[16]

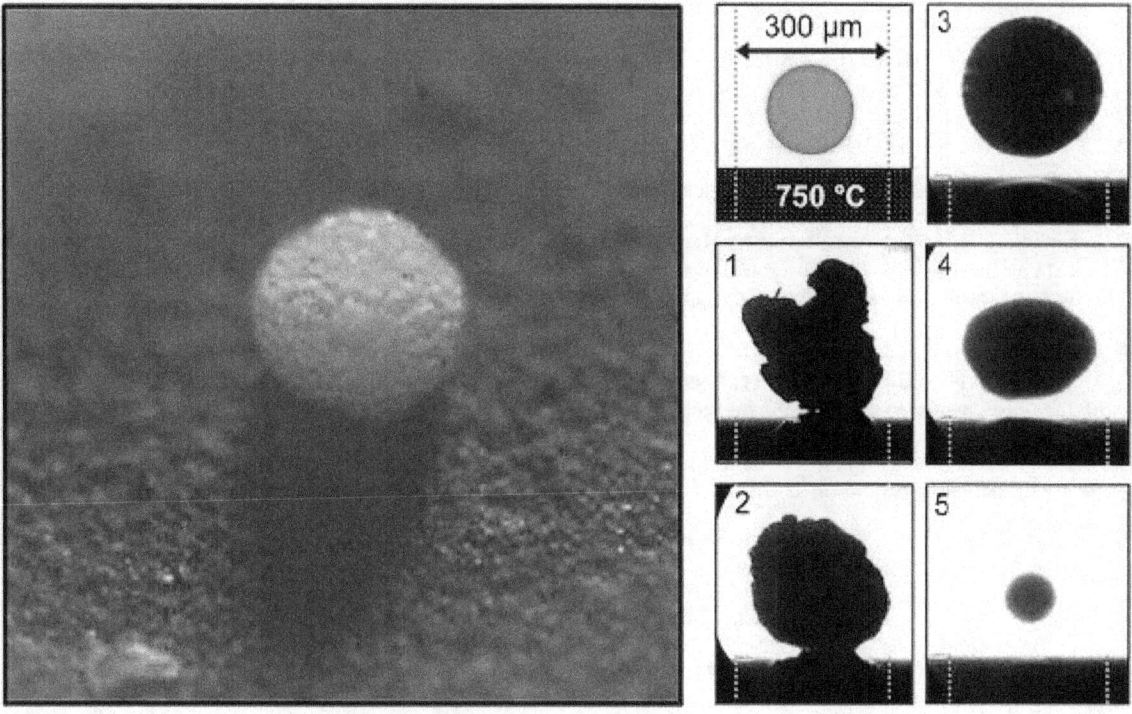

Reactive Leidenfrost effect of cellulose on silica, 750 °C (1,380 °F)

13.7 In popular culture

In Jules Verne's 1876 book *Michael Strogoff*, the protagonist is saved from being blinded with a hot blade by evaporating tears.

In the 2009 season finale of *MythBusters*, "Mini Myth Mayhem", the team demonstrated that a person can wet their hand and briefly dip it into molten lead without injury, using the Leidenfrost effect as the scientific basis.

In the fourth episode of the anime *Aldnoah.Zero*, the effect is mentioned as the cause of bullet tips melting midair after contacting plasma and subsequently being veered off course by the wind surrounding them.

13.8 See also

- Critical heat flux

- Mpemba effect

- Nucleate boiling

- region-beta paradox

13.9 References

[1] Willey, David (1999). "The Physics Behind Four Amazing Demonstrations". *Skeptical Inquirer* 23 (6). Retrieved 11 October 2014.

[2] Walker, Jearl. "Boiling and the Leidenfrost Effect" (PDF). *Fundamentals of Physics*: 1–4. Retrieved 11 October 2014.

[3] "Student Gulps Into Medical Literature". *Worcester Polytechnic Institute*. 20 January 1999. Retrieved 11 October 2014.

[4] Bernardin, John D.; Mudawar, Issam (2002). "A Cavity Activation and Bubble Growth Model of the Leidenfrost Point". *Journal of Heat Transfer* 124 (5): 864–74. doi:10.1115/1.1470487.

[5] Sir William Fairbairn (1851). *Two Lectures: The Construction of Boilers, and on Boiler Explosions, with the means of prevention*.

[6] Incropera, DeWitt, Bergman & Lavine: Fundamentals of Heat and Mass Transfer, 6th edition.

[7] Vakarelski, Ivan U.; Patankar, Neelesh A.; Marston, Jeremy O.; Chan, Derek Y. C.; Thoroddsen, Sigurdur T. (2012). "Stabilization of Leidenfrost vapour layer by textured superhydrophobic surfaces". *Nature* 489 (7415): 274–7. Bibcode:2012Natur.489..2 .doi:10.1038/nature11418. PMID 22972299.

[8] Subhrakanti Saha, Lee Chuin Chen, Mridul Kanti Mandal, Kenzo Hiraoka (March 2013). "Leidenfrost Phenomenon-assisted Thermal Desorption (LPTD) and Its Application to Open Ion Sources at Atmospheric Pressure Mass Spectrometry". *Journal of The American Society for Mass Spectrometry 3* (24): 341–7. doi:10.1007/s13361-012-0564-y.

[9] Wells, Gary G.; Ledesma-Aguilar, Rodrigio; McHale, Glen; Sefiane, Khellil (3 March 2015). "A sublimation heat engine". *Nature Communications*. Bibcode:2015NatCo...6E6390W. doi:10.1038/ncomms7390. Retrieved 5 March 2015.

[10] James R. Welty; Charles E. Wicks; Robert E. Wilson; Gregory L. Rorrer., "Fundamentals of Momentum, Heat and Mass transfer" 5th edition, John Wiley and Sons

[11] Carey, Van P. , Liquid Vapor Phase change Phenomena

[12] Berenson, P.J., *Film boiling heat transfer from a horizontal surface*, Journal of Heat Transfer, Volume 83, 1961, Pages 351-362

[13] Henry,R.E., *[A correlation for the minimum film boiling temperature]*,Chem. Eng. Prog. Symp. Ser. , Volume 70, 1974, Pages 81-90

[14] "Scientists levitate wood on structured surfaces captured by high speed photography" Phys.org. http://phys.org html

[15] "Reactive Liftoff of Crystalline Cellulose Particles", Scientific Reports 2015, 5, 11238. DOI: 10.1038/srep11238

[16] Scientists Levitate Wood on Structured Surfaces Captured by High Speed Photography. http://www.newswire.com/press-release/ scientists-levitate-wood-on-structured-surfaces-captured-by

13.10   External links

- Essay about the effect and demonstrations by Jearl Walker (PDF)

- Site with high-speed video, pictures and explanation of film-boiling by Heiner Linke at the University of Oregon, USA

- "Scientists make water run uphill" by BBC News about using the Leidenfrost effect for cooling of computer chips.

- "Uphill Water" - ABC Catalyst story

- "Leidenfrost Maze" - University of Bath undergraduate students Carmen Cheng and Matthew Guy

- "When Water Flows Uphill" - Science Friday with Univ of Bath professor Kei Takashina

- Jeffrey, Colin (March 10, 2015). "Engine running on frozen carbon dioxide may power mission to Mars". *Gizmag*. Retrieved March 2015.

Chapter 14

Macroscopic quantum phenomena

Quantum mechanics is most often used to describe matter on the scale of molecules, atoms, or elementary particles. However, some phenomena, particularly at low temperatures, show quantum behavior on a macroscopic scale. The best-known examples of **macroscopic quantum phenomena** are superfluidity and superconductivity; other examples include the quantum Hall effect and concerted proton tunneling in ice.[1] Since 2000 there has been extensive experimental work on quantum gases, particularly Bose–Einstein Condensates.

Between 1996 to 2003 four Nobel prizes were given for work related to macroscopic quantum phenomena.[2] Macroscopic quantum phenomena can be observed in superfluid helium and in superconductors,[3] but also in dilute quantum gases and in laser light. Although these media are very different, their behavior is very similar as they all show macroscopic quantum behavior.

Quantum phenomena are generally classified as macroscopic when the quantum states are occupied by a large number of particles (typically Avogadro's number) or the quantum states involved are macroscopic in size (up to km size in superconducting wires).[4]

14.1 Consequences of the macroscopic occupation

The concept of macroscopically-occupied quantum states is introduced by Fritz London.[5][6] In this section it will be explained what it means if the ground state is occupied by a very large number of particles. We start with the wave function of the ground state written as

with Ψ_0 the amplitude and φ the phase. The wave function is normalized so that

The physical interpretation of the quantity

depends on the number of particles. Fig.1 represents a container with a certain number of particles with a small control volume ΔV inside. We check from time to time how many particles are in the control box. We distinguish three cases:

1. There is only one particle. In this case the control volume is empty most of the time. However, there is a certain chance to find the particle in it given by Eq.(3). The chance is proportional to ΔV. The factor $\Psi\Psi^*$ is called the chance density.

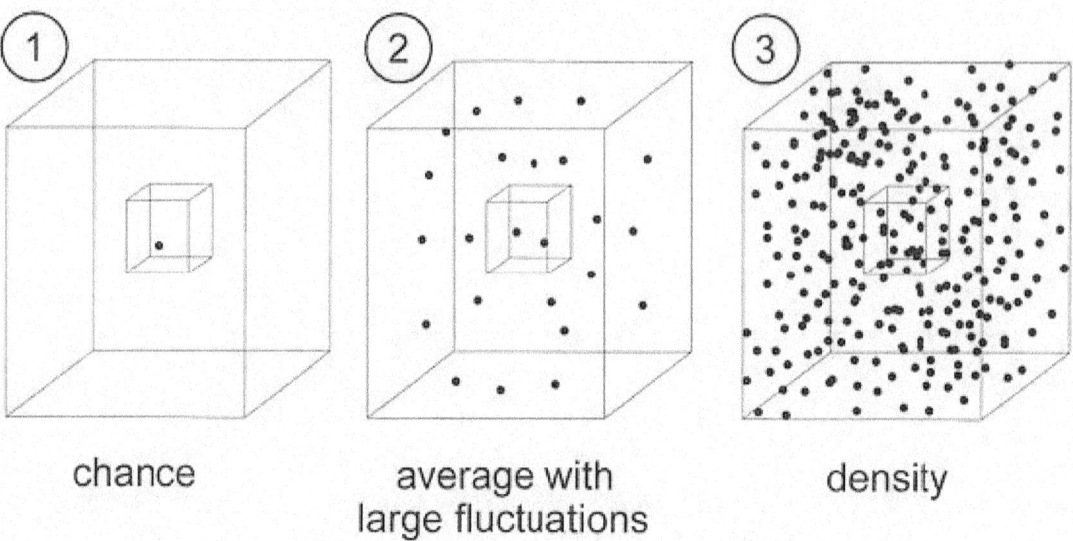

Fig.1 Left: only one particle; usually the small box is empty. However, there is a certain chance of that the particle is in the box. This chance is given by Eq.(15). Middle: a few particles. There are usually some particles in the box. We can define an average, but the actual number of particles in the box has large fluctuations around this average. Right: large number of particles. The fluctuations around the average are small.

2. If the number of particles is a bit larger there are usually some particles inside the box. We can define an average, but the actual number of particles in the box has relatively large fluctuations around this average.

3. In the case of a very large number of particles there will always be a lot of particles in the small box. The number will fluctuate but the fluctuations around the average are relatively small. The average number is proportional to ΔV and $\Psi\Psi^*$ is now interpreted as the particle density.

In quantum mechanics the particle probability flow density J_p (unit: particles per second per m^2) can be derived from the Schrödinger equation to be

with q the charge of the particle and $\vec{A}$ the vector potential. With Eq.(1)

If the wave function is macroscopically occupied the particle probability flow density becomes a particle flow density. We introduce the fluid velocity v_s via the mass flow density

The density (mass per m^3) is

so Eq.(5) results in

This important relation connects the velocity, a classical concept, of the condensate with the phase of the wave function, a quantum-mechanical concept.

14.2   Superfluidity

Main article: Superfluid
Below the lambda-temperature, helium shows the unique property of superfluidity. The fraction of the liquid that forms

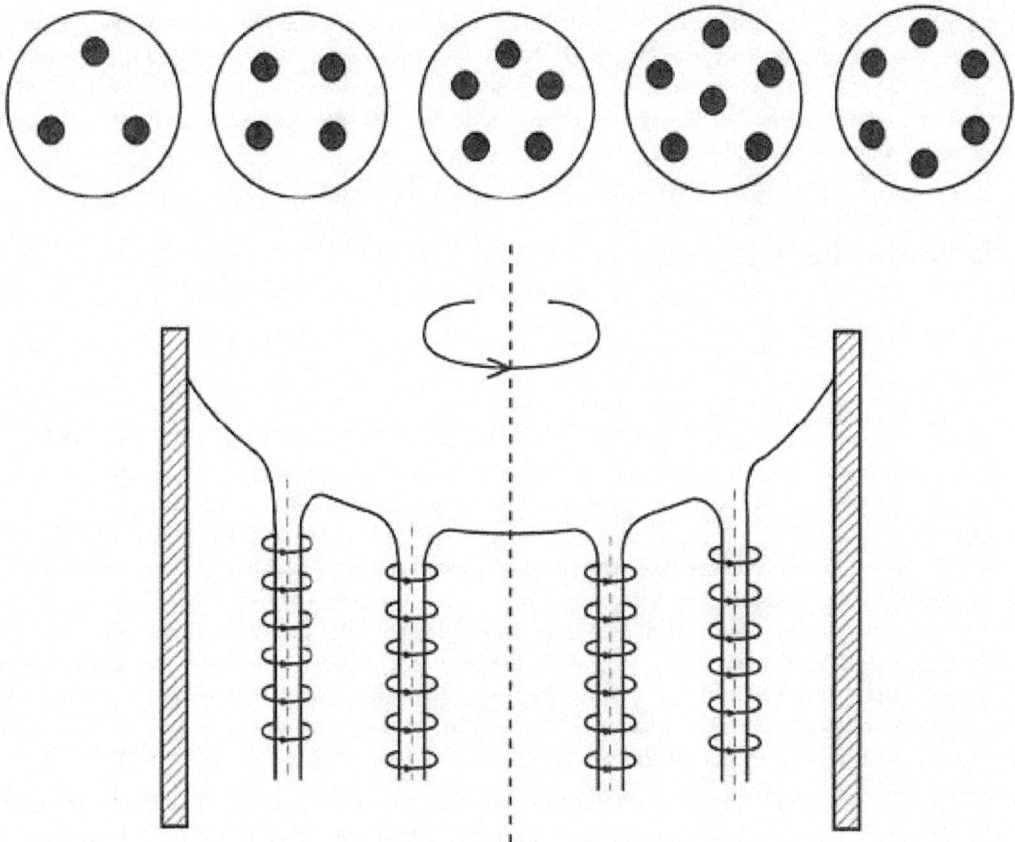

Fig.2 Lower part: vertical cross section of a column of superfluid helium rotating around a vertical axis. Upper part: Top view of the surface showing the pattern of vortex cores. From left to right the rotation speed is increased, resulting in an increasing vortex-line density.

the superfluid component is a macroscopic quantum fluid. The helium atom is a neutral particle, so $q=0$. Furthermore, when considering helium-4, the relevant particle mass is $m=m_4$, so Eq.(8) reduces to

For an arbitrary loop in the liquid, this gives

Due to the single-valued nature of the wave function

with n integer, we have

The quantity

is the quantum of circulation. For a circular motion with radius r

In case of a single quantum ($n=1$)

When superfluid helium is put in rotation, Eq.(13) will not be satisfied for all loops inside the liquid unless the rotation is organized around vortex lines (as depicted in Fig.2). These lines have a vacuum core with a diameter of about 1 Å (which is smaller than the average particle distance). The superfluid helium rotates around the core with very high speeds. Just outside the core ($r = 1$ Å), the velocity is as large as 160 m/s. The cores of the vortex lines and the container rotate as a solid body around the rotation axes with the same angular velocity. The number of vortex lines increases with the angular velocity (as shown in the upper half of the figure). Note that the two right figures both contain six vortex lines, but the lines are organized in different stable patterns.[7]

14.3   Superconductivity

Main article: Superconductivity

In the original paper Ginzburg and Landau observed the existence of two types of superconductors depending on the energy of the interface between the normal and superconducting states.The Meissner state breaks down when the applied magnetic field is too large. Superconductors can be divided into two classes according to how this breakdown occurs. In Type I superconductors, superconductivity is abruptly destroyed when the strength of the applied field rises above a critical value Hc. Depending on the geometry of the sample, one may obtain an intermediate state[8] consisting of a baroque pattern[9] of regions of normal material carrying a magnetic field mixed with regions of superconducting material containing no field. In Type II superconductors, raising the applied field past a critical value Hc_1 leads to a mixed state (also known as the vortex state) in which an increasing amount of magnetic flux penetrates the material, but there remains no resistance to the flow of electric current as long as the current is not too large. At a second critical field strength Hc_2, superconductivity is destroyed. The mixed state is actually caused by vortices in the electronic superfluid, sometimes called fluxons because the flux carried by these vortices is quantized. Most pure elemental superconductors, except niobium and carbon nanotubes, are Type I, while almost all impure and compound superconductors are Type II.

The most important finding from Ginzburg–Landau theory was made by Alexei Abrikosov in 1957. He used Ginzburg–Landau theory to explain experiments on superconducting alloys and thin films. He found that in a type-II superconductor in a high magnetic field, the field penetrates in a triangular lattice of quantized tubes of flux vortices.

14.3.1   Fluxoid quantization

For superconductors the bosons involved are the so-called Cooper pairs which are quasiparticles formed by two electrons.[10] Hence $m = 2m_e$ and $q = -2e$ where m_e and e are the mass of an electron and the elementary charge. It follows from Eq.(8) that

Integrating Eq.(15) over a closed loop gives

As in the case of helium we define the vortex strength

and use the general relation

where Φ is the magnetic flux enclosed by the loop. The so-called fluxoid is defined by

In general the values of κ and Φ depend on the choice of the loop. Due to the single-valued nature of the wave function and Eq.(16) the fluxoid is quantized

The unit of quantization is called the flux quantum

The flux quantum plays a very important role in superconductivity. The earth magnetic field is very small (about 50 μT), but it generates one flux quantum in an area of 6 by 6 μm. So, the flux quantum is very small. Yet it was measured to an accuracy of 9 digits as shown in Eq.(21). Nowadays the value given by Eq.(21) is exact by definition.

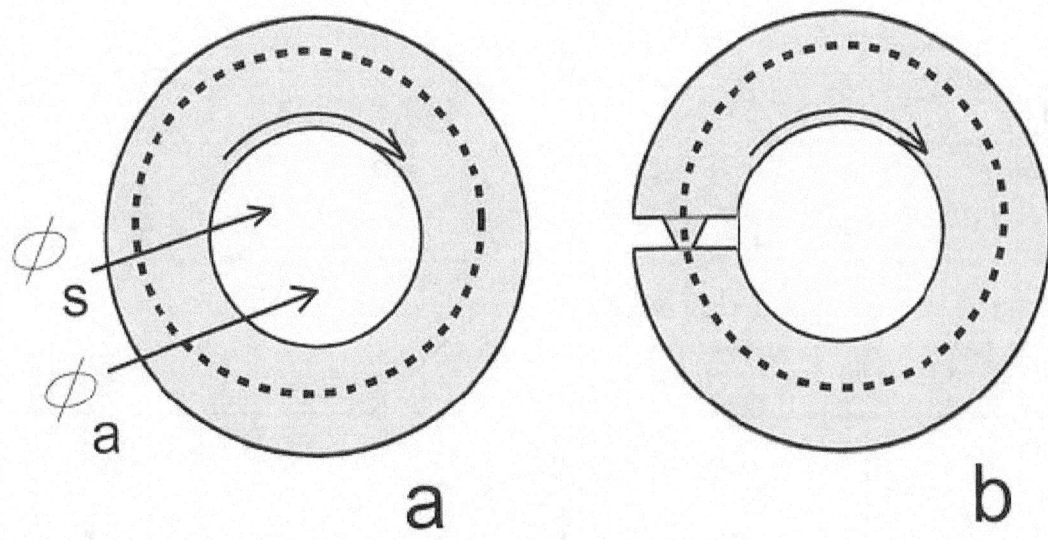

Fig. 3. Two superconducting rings in an applied magnetic field
a: thick superconducting ring. The integration loop is completely in the region with $v_s=0$;
b: thick superconducting ring with a weak link. The integration loop is completely in the region with $v_s=0$ except for a small region near the weak link.

In Fig. 3 two situations are depicted of superconducting rings in an external magnetic field. One case is a thick-walled ring and in the other case the ring is also thick-walled, but is interrupted by a weak link. In the latter case we will meet the famous Josephson relations. In both cases we consider a loop inside the material. In general a superconducting circulation current will flow in the material. The total magnetic flux in the loop is the sum of the applied flux Φ_a and the self-induced flux Φ_s induced by the circulation current

14.3.2   Thick ring

The first case is a thick ring in an external magnetic field (Fig. 3a). The currents in a superconductor only flow in a thin layer at the surface. The thickness of this layer is determined by the so-called London penetration depth. It is of μm size or less. We consider a loop far away from the surface so that $v_s=0$ everywhere so $\kappa=0$. In that case the fluxoid is equal to the magnetic flux ($\Phi_v=\Phi$). If $v_s=0$ Eq.(15) reduces to

Taking the rotation gives

Using the well-known relations $\vec{\nabla} \times \vec{\nabla}\varphi = 0$ and $\vec{\nabla} \times \vec{A} = \vec{B}$ shows that the magnetic field in the bulk of the superconductor is zero as well. So, for thick rings, the total magnetic flux in the loop is quantized according to

14.3.3   Interrupted ring, weak links

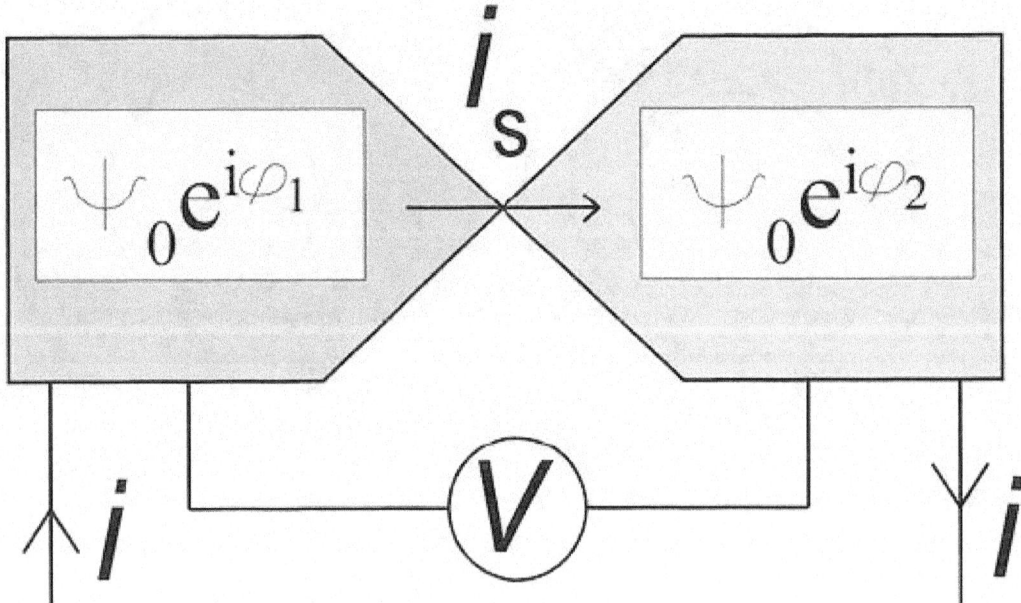

Fig. 4. Schematic of a weak link carrying a superconducting current is. The voltage difference over the link is V. The phases of the superconducting wave functions at the left and right side are assumed to be constant (in space, not in time) with values of φ_1 and φ_2 respectively.

Weak links play a very important role in modern superconductivity. In most cases weak links are oxide barriers between two superconducting thin films, but it can also be a crystal boundary (in the case of high-Tc superconductors). A schematic representation is given in Fig. 4. Now consider the ring which is thick everywhere except for a small section where the ring is closed via a weak link (Fig. 3b). The velocity is zero except near the weak link. In these regions the velocity contribution to the total phase change in the loop is given by (with Eq.(15))

The line integral is over the contact from one side to the other in such a way that the end points of the line are well inside the bulk of the superconductor where v_s=0. So the value of the line integral is well-defined (e.g. independent of the choice of the end points). With Eqs.(19), (22), and (26)

Without proof we state that the supercurrent through the weak link is given by the so-called DC Josephson relation[11]

The voltage over the contact is given by the AC Josephson relation

The names of these relations (DC and AC relations) are misleading since they both hold in DC and AC situations. In the steady state (constant $\Delta\varphi^*$) Eq.(29) shows that V=0 while a nonzero current flows through the junction. In the case of a constant applied voltage (voltage bias) Eq.(29) can be integrated easily and gives

Substitution in Eq.(28) gives

This is an AC current. The frequency

is called the Josephson frequency. One μV gives a frequency of about 500 MHz. By using Eq.(32) the flux quantum is determined with the high precision as given in Eq.(21).

The energy difference of a Cooper pair, moving from one side of the contact to the other, is $\Delta E = 2eV$. With this expression Eq.(32) can be written as $\Delta E = h\nu$ which is the relation for the energy of a photon with frequency ν.

The AC Josephson relation (Eq.(29)) can be easily understood in terms of Newton's law, (or from one of the London equation's[12]). We start with Newton's law

$$\vec{F} = m\mathrm{d}\vec{v}_s/\mathrm{d}t.$$

Substituting the expression for the Lorentz force

$$\vec{F} = q(\vec{E} + \vec{v}_s \times \vec{B})$$

and using the general expression for the co-moving time derivative

$$d\vec{v}_s/dt = \partial\vec{v}_s/\partial t + (1/2)\vec{\nabla}v_s^2 - \vec{v}_s \times (\vec{\nabla} \times \vec{v}_s)$$

gives

$$(q/m)(\vec{E} + \vec{v}_s \times \vec{B}) = \partial\vec{v}_s/\partial t + (1/2)\vec{\nabla}v_s^2 - \vec{v}_s \times (\vec{\nabla} \times \vec{v}_s).$$

Eq.(8) gives

$$0 = \vec{\nabla} \times \vec{v}_s + (q/m)\vec{\nabla} \times \vec{A} = \vec{\nabla} \times \vec{v}_s + (q/m)\vec{B}$$

so

$$(q/m)\vec{E} = \partial\vec{v}_s/\partial t + (1/2)\vec{\nabla}v_s^2.$$

Take the line integral of this expression. In the end points the velocities are zero so the ∇v^2 term gives no contribution. Using

$$\int \vec{E} \cdot \vec{dl} = -V$$

and Eq.(26), with $q = -2e$ and $m = 2m_e$, gives Eq.(29).

14.3.4   DC SQUID

Main article: SQUID

Figure 5 shows a so-called DC SQUID. It consists of two superconductors connected by two weak links. The fluxoid quantization of a loop through the two bulk superconductors and the two weak links demands

If the self-inductance of the loop can be neglected the magnetic flux in the loop Φ is equal to the applied flux

with B the magnetic field, applied perpendicular to the surface, and A the surface area of the loop. The total supercurrent is given by

Substitution of Eq(33) in (35) gives

Using a well known geometrical formula we get

Since the sin-function can vary only between -1 and $+1$ a steady solution is only possible if the applied current is below a critical current given by

Note that the critical current is periodic in the applied flux with period Φ_0. The dependence of the critical current on the applied flux is depicted in Fig. 6. It has a strong resemblance with the interference pattern generated by a laser beam behind a double slit. In practice the critical current is not zero at half integer values of the flux quantum of the applied flux. This is due to the fact that the self-inductance of the loop cannot be neglected.[13]

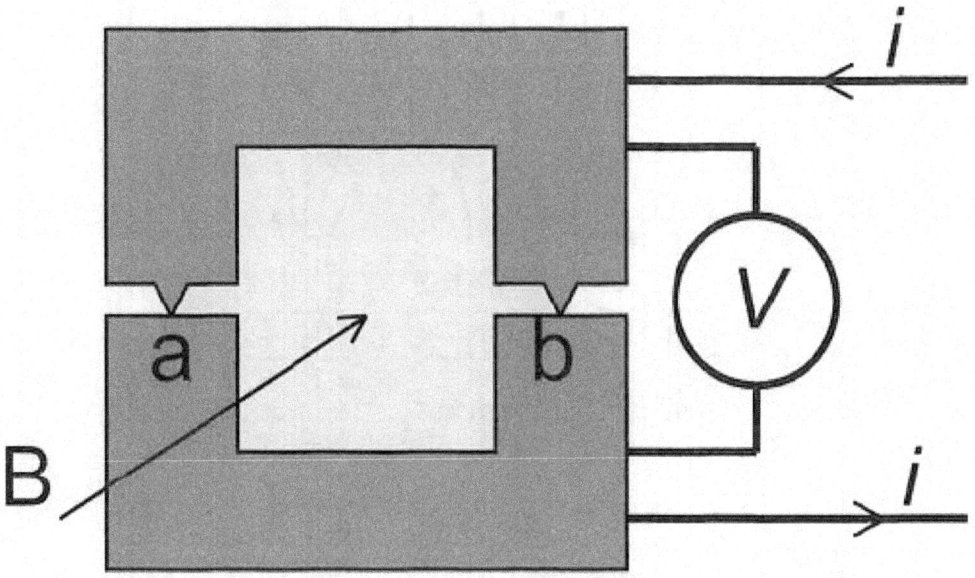

Fig. 5. *Two superconductors connected by two weak links. A current and a magnetic field are applied.*

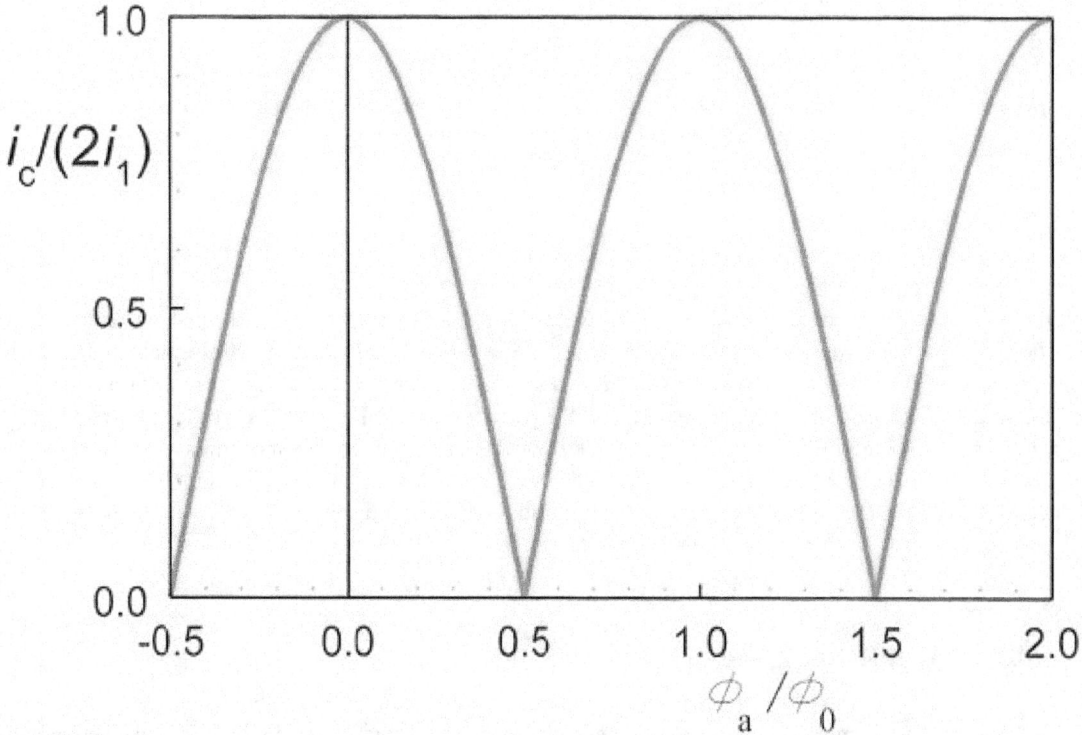

Fig. 6. *Dependence of the critical current of a DC-SQUID on the applied magnetic field*

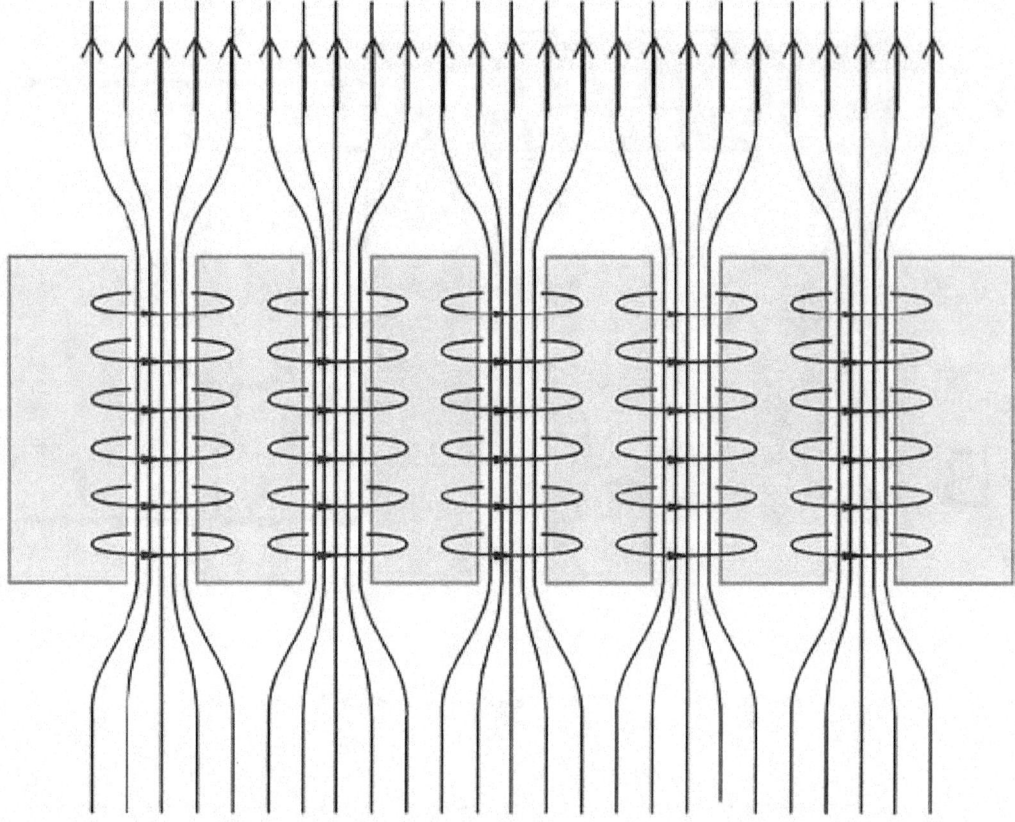

Fig. 7. Magnetic flux lines penetrating a type-II superconductor. The currents in the superconducting material generate a magnetic field which, together with the applied field, result in bundles of quantized flux.

14.3.5 Type II superconductivity

Main article: Type-II superconductor

  Type-II superconductivity is characterized by two critical fields called B_{c1} and B_{c2}. At a magnetic field B_{c1} the applied magnetic field starts to penetrate the sample, but the sample is still superconducting. Only at a field of B_{c2} the sample is completely normal. For fields in between B_{c1} and B_{c2} magnetic flux penetrates the superconductor in well-organized patterns, the so-called Abrikosov vortex lattice similar to the pattern shown in Fig. 2.[14] A cross section of the superconducting plate is given in Fig. 7. Far away from the plate the field is homogeneous, but in the material superconducting currents flow which squeeze the field in bundles of exactly one flux quantum. The typical field in the core is as big as 1 tesla. The currents around the vortex core flow in a layer of about 50 nm with current densities on the order of 15×10^{12} A/m^2. That corresponds with 15 million ampère in a wire of one mm^2.

14.4 Dilute quantum gases

The classical types of quantum systems, superconductors and superfluid helium, were discovered in the beginning of the 20th century. Near the end of the 20th century, scientists discovered how to create very dilute atomic or molecular gases, cooled first by laser cooling and then by evaporative cooling.[15] They are trapped using magnetic fields or optical dipole potentials in ultrahigh vacuum chambers. Isotopes which have been used include rubidium (Rb-87 and Rb-85), strontium (Sr-87, Sr-86, and Sr-84) potassium (K-39 and K-40), sodium (Na-23), lithium (Li-7 and Li-6), and hydrogen (H-1). The temperatures to which they can be cooled are as low as a few nanokelvin. The developments have been very fast in the past few years. A team of NIST and the University of Colorado has succeeded in creating and observing vortex quantization in these systems.[16] The concentration of vortices increases with the angular velocity of the rotation, similar to the case of superfluid helium and superconductivity.

14.5   See also

- Quantum turbulence

- Schrödinger's cat paradox

- Second sound

- Flux quantization

- Domain wall (magnetism)

- Flux pinning

- Ginzburg–Landau theory

- Husimi Q representation

- Magnetic flux quantum

- Quantum vortex

- Topological defect

- Superconductivity

- Type-I superconductor

- Type-II superconductor

- Meissner effect

- SQUID

- Josephson effect

- Charge density wave

14.6   References and footnotes

[1] Yen, F., and Gao, T. (June 2015). "Dielectric Anomaly in Ice near 20 K; Evidence of Macroscopic Quantum Phenomena". *The Journal of Physical Chemistry Letters* **6** (14): 2822–2825. doi:10.1021/acs.jpclett.5b00797.

[2] These Nobel prizes were for the discovery of super-fluidity in helium-3 (1996), for the discovery of the fractional quantum Hall effect (1998), for the demonstration of Bose–Einstein condensation (2001), and for contributions to the theory of superconductivity and superfluidity (2003).

[3] D.R. Tilley and J. Tilley, *Superfluidity and Superconductivity*, Adam Hilger, Bristol and New York, 1990

[4] Jaeger, Gregg (September 2014). "What in the (quantum) world is macroscopic?". *American Journal of Physics* **82** (9): 896–905. Bibcode:2014AmJPh..82..896J. doi:10.1119/1.4878358.

[5] Fritz London *Superfluids* (London, Wiley, 1954-1964)

[6] Gavroglu, K.; Goudaroulis, Y. (1988). "Understanding macroscopic quantum phenomena: The history of superfluidity 1941–1955". *Annals of Science* **45** (4): 367. doi:10.1080/00033798800200291.

[7] E.J. Yarmchuk and R.E. Packard (1982). "Photographic studies of quantized vortex lines". *J. Low Temp. Phys.* **46** (5–6): 479. Bibcode:1982JLTP...46..479Y. doi:10.1007/BF00683912.

[8] Lev D. Landau; Evgeny M. Lifschitz (1984). *Electrodynamics of Continuous Media*. Course of Theoretical Physics **8**. Oxford: Butterworth-Heinemann. ISBN 0-7506-2634-8.

[9] David J. E. Callaway (1990). "On the remarkable structure of the superconducting intermediate state". *Nuclear Physics B* **344** (3): 627–645. Bibcode:1990NuPhB.344..627C. doi:10.1016/0550-3213(90)90672-Z.

[10] M. Tinkham (1975). *Introduction to Superconductivity*. McGraw-Hill.

[11] B.D. Josephson (1962). "Possible new eff ects in superconductive tunneling". *Phys. Lett.* **1** (7): 251–253. Bibcode: 1962PhLt...doi:10.1016/0031-9163(62)91369-0.

[12] London, F.; London, H. (1935). "The Electromagnetic Equations of the Supraconductor". *Proceedings of the Royal Society A: Mathematical, Physical and Engineering Sciences* **149** (866): 71. Bibcode:1935RSPSA.149...71L. doi:10.1098/rspa.1935.0048.

[13] A.TH.A.M. de Waele and R. de Bruyn Ouboter (1969). "Quantum-interference phenomena in point contacts between two superconductors". *Physica* **41** (2): 225–254. Bibcode:1969Phy....41..225D. doi:10.1016/0031-8914(69)90116-5.

[14] Essmann, U.; Träuble, H. (1967). "The direct observation of individual flux lines in type II superconductors". *Physics Letters A* **24** (10): 526. Bibcode:1967PhLA...24..526E. doi:10.1016/0375-9601(67)90819-5.

[15] Anderson, M.H., Ensher, J.R., Matthews, M.R., Wieman, C.E., and Cornell, E.A. (1995). "Observation of Bose-Einstein Condensation in a Dilute Atomic Vapor". *Science* **269** (5221): 198–201. Bibcode:1995Sci...269..198A. doi:10.1126/science.269.5 221.198.PMID 17789847.

[16] Schweikhard, V., Coddington, I., Engels, P., Tung, S., and Cornell, E.A. (2004). "Vortex-Lattice Dynamics in Rotating Spinor Bose-Einstein Condensates". *Phys. Rev. Lett.* **93** (3): 210403. Bibcode:2004PhRvL..93c0403N. doi:10.1103/PhysRevLett.93 .030403.

Chapter 15

Mpemba effect

The **Mpemba effect**, named after Erasto Mpemba in 1969, is the observation that, in some circumstances, warmer water can freeze faster than colder water. Although there is evidence of the effect, there is disagreement on exactly what the effect is and under what circumstances it occurs. However, in 1966 Keith Houser, a Senior at Port Washington High School in Ohio entered the Ohio State Science Fair proving warmer water he used in the Winter for cattle froze quicker than cold water. There have been reports of similar phenomena since ancient times, although with insufficient detail for the claims to be replicated. A number of possible explanations for the effect have been proposed. Further investigations will need to decide on a precise definition of "freezing" and control a vast number of starting parameters in order to confirm or explain the effect.

15.1 Definition

The phenomenon, when taken to mean "hot water freezes faster than cold", is difficult to reproduce or confirm, because this statement is ill defined.[1] Jeng proposes as a more precise wording:

> There exists a set of initial parameters, and a pair of temperatures, such that given two bodies of water identical in these parameters, and differing only in initial uniform temperatures, the hot one will freeze sooner.[2]

However, even with this definition it is not clear whether "freezing" refers to the point at which water forms a visible surface layer of ice; the point at which the entire volume of water becomes a solid block of ice; or when the water reaches 0 °C (32 °F).[1] A quantity of water can be at 0 °C (32 °F) and not be ice; after enough calories (i.e., heat) have been removed to reach 0 °C (32 °F) more calories must be removed before the water changes to solid state (ice), so water can be liquid or solid at 0 °C (32 °F).[3]

With the above definition there are simple ways in which the effect might be observed: For example, if the hotter temperature melts the frost on a cooling surface and thus increases the thermal conductivity between the cooling surface and the water container.[1] On the other hand, there may be many circumstances in which the effect is not observed.[1]

15.2 Observations

15.2.1 Historical context

Various effects of heat on the freezing of water were described by ancient scientists such as Aristotle: "The fact that the water has previously been warmed contributes to its freezing quickly: for so it cools sooner. Hence many people, when they want to cool water quickly, begin by putting it in the sun. So the inhabitants of Pontus when they encamp on the ice to fish (they cut a hole in the ice and then fish) pour warm water round their reeds that it may freeze the quicker, for they use the ice like lead to fix the reeds."[4] Aristotle's explanation involved *antiperistasis*, "the supposed increase in the intensity of a quality as a result of being surrounded by its contrary quality."

Early modern scientists such as Francis Bacon noted that "slightly tepid water freezes more easily than that which is utterly cold."[5] In the original Latin, "aqua parum tepida facilius conglacietur quam omnino frigida."

René Descartes wrote in his *Discourse on the Method*, "One can see by experience that water that has been kept on a fire for a long time freezes faster than other, the reason being that those of its particles that are least able to stop bending evaporate while the water is being heated."[6] This relates to Descartes' vortex theory.

James Black investigated a special case of this phenomenon comparing previously-boiled with unboiled water;[7] the previously-boiled water froze more quickly. Evaporation was controlled for. He discussed the influence of stirring on the results of the experiment, noting that stirring the unboiled water led to it freezing at the same time as the previously-boiled water, and also noted that stirring the very-cold unboiled water led to immediate freezing. James Black then discussed this Fahrenheit's description of supercooling of water (although the term supercooling had not then been coined), arguing, in modern terms, that the previously-boiled water could not be as readily supercooled.

15.2.2 Mpemba's observation

The effect is named after Tanzanian Erasto Mpemba. He described in 1963 in Form 3 of Magamba Secondary School, Tanganyika, when freezing ice cream mix that was hot in cookery classes and noticing that it froze before the cold mix. He later became a student at Mkwawa Secondary (formerly High) School in Iringa. The headmaster invited Dr. Denis G. Osborne from the University College in Dar es Salaam to give a lecture on physics. After the lecture, Erasto Mpemba asked him the question "If you take two similar containers with equal volumes of water, one at 35 °C (95 °F) and the other at 100 °C (212 °F), and put them into a freezer, the one that started at 100 °C (212 °F) freezes first. Why?", only to be ridiculed by his classmates and teacher. After initial consternation, Osborne experimented on the issue back at his workplace and confirmed Mpemba's finding. They published the results together in 1969, while Mpemba was studying at the College of African Wildlife Management.[8]

15.2.3 Modern context

Mpemba and Osborne describe placing 70 ml (2.5 imp fl oz; 2.4 US fl oz) samples of water in 100 ml (3.5 imp fl oz; 3.4 US fl oz) beakers in the ice box of a domestic refrigerator on a sheet of polystyrene foam. They showed the time for *freezing to start* was longest with an initial temperature of 25 °C (77 °F) and that it was much less at around 90 °C (194 °F). They ruled out loss of liquid volume by evaporation as a significant factor and the effect of dissolved air. In their setup most heat loss was found to be from the liquid surface.[8]

David Auerbach describes an effect that he observed in samples in glass beakers placed into a liquid cooling bath. In all cases the water supercooled, reaching a temperature of typically −6 to −18 °C (21 to 0 °F) before spontaneously freezing. Considerable random variation was observed in the time required for spontaneous freezing to start and in some cases this resulted in the water which started off hotter (partially) freezing first.[9]

In studies appearing in Phys.org, James Brownridge, a radiation safety officer at the State University of New York, indicates supercooling is involved.[10]

15.3 Suggested explanations

The behaviour seems contrary to natural expectation but many explanations have been proposed.

- *Evaporation*: The evaporation of the warmer water reduces the mass of the water to be frozen.[11] Evaporation is endothermic, meaning that the water mass is cooled by vapor carrying away the heat, but this alone probably does not account for the entirety of the effect.[2]

- *Convection*: Accelerating heat transfers. Reduction of water density below 4 °C (39 °F) tends to suppress the convection currents that cool the lower part of the liquid mass; the lower density of hot water would reduce this effect, perhaps sustaining the more rapid initial cooling. Higher convection in the warmer water may also spread ice crystals around faster.[12]

- *Frost*: Has insulating effects. The lower temperature water will tend to freeze from the top, reducing further heat loss by radiation and air convection, while the warmer water will tend to freeze from the bottom and sides because of water convection. This is disputed as there are experiments that account for this factor.[2]

- *Supercooling*: It is hypothesised that cold water, when placed in a freezing environment, supercools more than hot water in the same environment, thus solidifying slower than hot water.[13][14] However, super-cooling tends to be less significant where there are particles that act as nuclei for ice crystals, thus precipitating rapid freezing.

- *Solutes*: The effects of calcium carbonate, magnesium carbonate among others.[15]

- *Thermal conductivity*: The container of hotter liquid may melt through a layer of frost that is acting as an insulator under the container (frost is an insulator, as mentioned above), allowing the container to come into direct contact with a much colder lower layer that the frost formed on (ice, refrigeration coils, etc.) The container now rests on a much colder surface (or one better at removing heat, such as refrigeration coils) than the originally colder water, and so cools far faster from this point on.

- *Dissolved Gases*: Cold water can contain more dissolved gases than hot water, which may somehow change the properties of the water with respect to convection currents, a proposition that has some experimental support but no theoretical explanation.[2]

15.4   Recent views

A reviewer for *Physics World* writes, "Even if the Mpemba effect is real — if hot water can sometimes freeze more quickly than cold — it is not clear whether the explanation would be trivial or illuminating." He pointed out that investigations of the phenomenon need to control a large number of initial parameters (including type and initial temperature of the water, dissolved gas and other impurities, and size, shape and material of the container, and temperature of the refrigerator) and need to settle on a particular method of establishing the time of freezing, all of which might affect the presence or absence of the Mpemba effect. The required vast multidimensional array of experiments might explain why the effect is not yet understood.[1]

New Scientist recommends starting the experiment with containers at 35 and 5 °C (95 and 41 °F) to maximize the effect.[16] In a related study, it was found that freezer temperature also affects the probability of observing the Mpemba phenomena as well as container temperature. For a liquid bath freezer, a temperature range of −3 to −8 °C (27 to 18 °F) was recommended.[14]

In 2012, the Royal Society of Chemistry held a competition calling for papers offering explanations to the Mpemba effect.[17] More than 22,000 people entered and Erasto Mpemba himself announced Nikola Bregović as the winner, suggesting that convection and supercooling were the reasons for the effect.[18]

15.5   See also

Other phenomena in which large effects may be achieved faster than small effects are

- Latent heat: turning 0 °C (32 °F) water to 0 °C (32 °F) ice takes the same amount of energy from the water as cooling it from 80 °C (176 °F) to 0 °C (32 °F)

- Leidenfrost effect: lower temperature boilers can sometimes vaporize water faster than higher temperature boilers

- Region-beta paradox: people can sometimes recover more quickly from more intense emotions or pain than from less distressing experiences

- Water cluster

15.6   References

[1] Ball, Philip (April 2006). Does hot water freeze first?. *Physics World*, pp. 19-26.

[2] Jeng, Monwhea (2006). "Hot water can freeze faster than cold?!?". *American Journal o f Physics* 74 (6): 514. Bibcode:2006AmJPh..74..514J. doi:10.1119/1.2186331.

[3] High School Chemistry and Physics

[4] Aristotle, Meteorology I.12 348b31–349a4

[5] Francis Bacon, Novum Organum, Lib. II, L

[6] Descartes, Les Meteores

[7] Black, James (1 January 1775). "The Supposed Effect of Boiling upon Water, in Disposing It to Freeze More Readily, Ascertained by Experiments. By Joseph Black, M. D. Professor of Chemistry at Edinburgh, in a Letter to Sir John Pringle, Bart. P. R. S.". *Philosophical Transactions of the Royal Society of London* 65: 124–128. doi:10.1098/rstl.1775.0014.

[8] Mpemba, Erasto B.; Osborne, Denis G. (1969). "Cool?". *Physics Education* (Institute of Physics) 4: 172–175 doi:10.1088/0031-9120/4/3/312. republished as Mpemba, E B; Osborne, D G (1979). "The Mpemba effect" (PDF). *Physics Education* (Institute of Physics) 14: 410–412. Bibcode:1979PhyEd..14..410M. doi:10.1088/0031-9120/14/7/312.

[9] Auerbach, David (1995). "Supercooling and the Mpemba effect: when hot water freezes quicker than cold" (PDF). *American Journal of Physics* 63 (10): 882–885. Bibcode:1995AmJPh..63..882A. doi:10.1119/1.18059.

[10] Edwards, Lin (26 March 2010). "Mpemba effect: Why hot water can freeze faster than cold". SUNY: Science X Network, Phys.org.

[11] Kell, G. S. (1969). "The freezing of hot and cold water". *Am. J. Phys.* 37 (5): 564–565. Bibcode:1969AmJPh..37..564K. doi:10.1119/1.1975687.

[12] CITV *Prove It!* Series 1 Programme 13

[13] S. Esposito, R. De Risi and L. Somma (2008). "Mpemba effect and phase transitions in the adiabatic cooling of water before freezing". *Physica A* 387 (4): 757–763. arXiv:0704.1381. Bibcode:2008PhyA..387..757E. doi:10.1016/j.physa.2007.10.029.

[14] Gholaminejad, Amir; Reza Hosseini (March 2013). "A Study of Water Supercooling". *Journal of Electronics Cooling and Thermal Control* 3: 1–6. Bibcode:2013JECTC...3....1G. doi:10.4236/jectc.2013.31001.

[15] Katz, Jonathan (April 2006). "When hot water freezes before cold". arXiv:physics/0604224 [physics.chem-ph].

[16] How to Fossilize Your Hamster: And Other Amazing Experiments For The Armchair Scientist, ISBN 1-84668-044-1

[17] Mpemba Competition. Royal Society of Chemistry. 2012.

[18] Winner of the Mpemba Competition. Royal Society of Chemistry. 2013.

15.7   Bibliography

- Dorsey, N. Ernest (1948). "The freezing of supercooled water". *Trans. Am. Phil. Soc.* (American Philosophical Society) 38 (3): 247–326. doi:10.2307/1005602. JSTOR 1005602. An extensive study of freezing experiments.

- Auerbach, David (1995). "Supercooling and the Mpemba effect: when hot water freezes quicker than cold" (PDF). *American Journal of Physics* 63 (10): 882–885. Bibcode:1995AmJPh..63..882A. doi:10.1119/1.18059. Auerbach attributes the Mpemba effect to differences in the behaviour of supercooled formerly hot water and formerly cold water.

- Knight, Charles A. (May 1996). "The MPEMBA effect: The freezing times of hot and cold water". *American Journal of Physics* **64** (5): 524. Bibcode:1996AmJPh..64..524K. doi:10.1119/1.18275.

- Monwhea, Jeng (2006). "The Mpemba effect: When can hot water freeze faster than cold?". *American Journal of Physics* **74** (6): 514. arXiv:physics/0512262. Bibcode:2006AmJPh..74..514J. doi:10.1119/1.2186331.

- Chown, Marcus (June 2006). "Why water freezes faster after heating". *New scientist*.

15.8 External links

- Mpemba Competition - Royal Society of Chemistry

- Xi Zhang Yongli Huang, Zengsheng Ma, Chang Q Sun. "O:H-O Bond Anomalous Relaxation Resolving Mpemba Paradox".

- Mpemba, E B; Osborne, D G. "The Mpemba effect" (PDF). Institute of Physics.

- Adams, Cecil; Mary M.Q.C. (1996). "Which freezes faster, hot water or cold water?". *The Straight Dope*. Chicago Reader, Inc. Retrieved January 2008.

- Brownridge, James (2010). "A search for the Mpemba effect: When hot water freezes faster than cold water". arXiv:1003.3185 [physics.pop-ph].

- "Heat questions". *HyperPhysics*. Georgia State University.

- "The Mpemba Effect". - History and analysis of the Mpemba Effect.

- Jeng, Monwhea (November 1998). "Can hot water freeze faster than cold water?". in the University of California Usenet Physics FAQ

- "The Phase Anomalies of Water: Hot Water may Freeze Faster than Cold Water". An analysis of the Mpemba effect. London South Bank University.

- "Mpemba effect: Why hot water can freeze faster than cold". A possible explanation of the Mpemba Effect

- "The story of the Mpemba effect told by the protagonists". An historical interview with Erasto Mpemba, Dr Denis Osborne and Ray deSouza

Chapter 16

Order and disorder (physics)

In physics, the terms **order** and **disorder** designate the presence or absence of some symmetry or correlation in a many-particle system.

In condensed matter physics, systems typically are ordered at low temperatures; upon heating, they undergo one or several phase transitions into less ordered states. Examples for such an **order-disorder transition** are:

- the melting of ice: solid-liquid transition, loss of crystalline order;

- the demagnetization of iron by heating above the Curie temperature: ferromagnetic-paramagnetic transition, loss of magnetic order.

The degree of freedom that is ordered or disordered can be translational (crystalline ordering), rotational (ferroelectric ordering), or a spin state (magnetic ordering).

The order can consist either in a full crystalline space group symmetry, or in a correlation. Depending on how the correlations decay with distance, one speaks of **long-range order** or **short-range order**.

If a disordered state is not in thermodynamic equilibrium, one speaks of **quenched disorder**. For instance, a glass is obtained by quenching (supercooling) a liquid. By extension, other quenched states are called spin glass, orientational glass. In some contexts, the opposite of quenched disorder is **annealed disorder**.

16.1 Characterizing order

16.1.1 Lattice periodicity and X-ray crystallinity

The strictest form of order in a solid is **lattice periodicity**: a certain pattern (the arrangement of atoms in a unit cell) is repeated again and again to form a translationally invariant tiling of space. This is the defining property of a crystal. Possible symmetries have been classified in 14 Bravais lattices and 230 space groups.

Lattice periodicity implies **long-range order**: if only one unit cell is known, then by virtue of the translational symmetry it is possible to accurately predict all atomic positions at arbitrary distances. During much of the 20th century, the converse was also taken for granted - until the discovery of quasicrystals in 1982 showed that there are perfectly deterministic tilings that do not possess lattice periodicity.

Besides structural order, one may consider charge ordering, spin ordering, magnetic ordering, and compositional ordering. Magnetic ordering is observable in neutron diffraction.

It is a thermodynamic entropy concept often displayed by a second-order phase transition. Generally speaking, high thermal energy is associated with disorder and low thermal energy with ordering, although there have been violations of this. Ordering peaks become apparent in diffraction experiments at low angles.

16.1.2 Long-range order

Long-range order characterizes physical systems in which remote portions of the same sample exhibit correlated behavior.

This can be expressed as a correlation function, namely the spin-spin correlation function:

$$G(x, x') = \langle s(x), s(x') \rangle.$$

where s is the spin quantum number and x is the distance function within the particular system.

This function is equal to unity when $x = x'$ and decreases as the distance $|x - x'|$ increases. Typically, it decays exponentially to zero at large distances, and the system is considered to be disordered. If, however, the correlation function decays to a constant value at large $|x - x'|$ then the system is said to possess long-range order. If it decays to zero as a power of the distance then it is called quasi-long-range order (for details see Chapter 11 in the textbook cited below. See also Berezinskii–Kosterlitz–Thouless transition). Note that what constitutes a large value of $|x - x'|$ is relative.

16.2 Quenched disorder

In statistical physics, a system is said to present **quenched disorder** when some parameters defining its behaviour are random variables which do not evolve with time, i.e.: they are quenched or *frozen*. Spin glasses are a typical example. It is opposite to annealed disorder, where the random variables are allowed to evolve themselves.

In mathematical terms, quenched disorder is harder to analyze than its annealed counterpart, since the thermal and the noise averaging play very different roles. In fact, the problem is so hard that few techniques to approach each are known, most of them relying on approximations. The most used are 1) a technique based on a mathematical analytical continuation known as the replica trick and 2) the Cavity method; although these give results in accord with experiments in a large range of problems, they are not generally proven to be a rigorous mathematical procedure. More recently it has been shown by rigorous methods, however, that at least in the archetypal spin-glass model (the so-called Sherrington-Kirkpatrick model) the replica based solution is indeed exact. The second most used technique in this field is generating functional analysis. This method is based on path integrals, and is in principle fully exact, although generally more difficult to apply than the replica procedure.

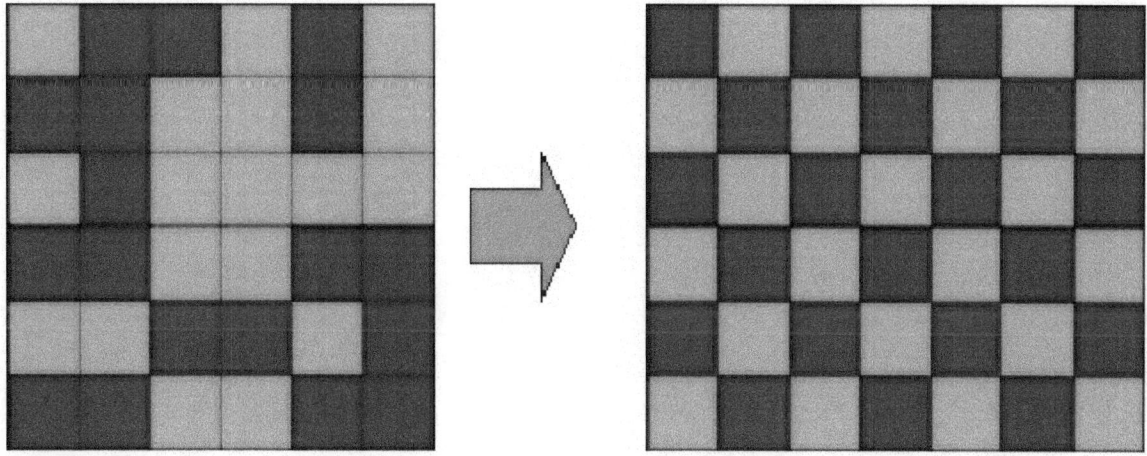

Transition from disordered (left) to ordered (right) states

16.3   Annealed disorder

A system is said to present **annealed disorder** when some parameters entering its definition are random variables, but whose evolution is related to that of the degrees of freedom defining the system. It is defined in opposition to quenched disorder, where the random variables may not change its value.

Systems with annealed disorder are usually considered to be easier to deal with mathematically, since the average on the disorder and the thermal average may be treated on the same footing.

16.4   See also

- In high energy physics, the formation of the chiral condensate in quantum chromodynamics is an ordering transition; it is discussed in terms of superselection.

- Entropy

- Topological order

- Impurity

- superstructure (physics)

16.5   Further reading

- H Kleinert: *Gauge Fields in Condensed Matter* (ISBN 9971-5-0210-0, 2 volumes) Singapore: World Scientific (1989).

Chapter 17

Spinodal

In thermodynamics, the limit of local stability with respect to small fluctuations is clearly defined by the condition that the second derivative of Gibbs free energy is zero. The locus of these points (the inflection point within a G-x or G-c curve, Gibbs free energy as a function of composition) is known as the **spinodal** curve.[1][2] For compositions within this curve, infinitesimally small fluctuations in composition and density will lead to phase separation via spinodal decomposition. Outside of the curve, the solution will be at least metastable with respect to fluctuations.[2] In other words, outside the spinodal curve some careful process may obtain a single phase system.[2] Inside it, only processes far from thermodynamic equilibrium, such as physical vapor deposition, will allow to prepare single phase compositions.[3] The local points of coexisting compositions, defined by the common tangent construction, are known as binodal (coexistence) curve, which denotes the minimum-energy equilibrium state of the system. Increasing temperature results in a decreasing difference between mixing entropy and mixing enthalpy, thus, the coexisting compositions come closer. The binodal curve forms the bases for the miscibility gap in a phase diagram. Since the free energy of mixture changes with temperature and concentration, the binodal and spinodal meet at the critical or consulate temperature and composition.[4]

17.1   Criterion

For binary solutions, the thermodynamic criterion which defines the spinodal curve is that the second derivative of free energy with respect to density or some composition variable is zero.[2][5][6]

17.2   Critical point

Extrema of the spinodal in temperature with respect to the composition variable coincide with ones of the binodal curve and are known as critical points.[6]

17.3   References

[1] Sandler S. I., *Chemical and Engineering Thermodynamics*. 1999 John Wiley & Sons, Inc., p 571.

[2] Koningsveld K., Stockmayer W. H.,Nies, E., *Polymer Phase Diagrams: A Textbook*. 2001 Oxford, p 12.

[3] P.H. Mayrhofer et al. Progress in Materials Science 51 (2006) 1032-1114 doi:10.1016/j.pmatsci.2006.02.002

[4] Cahn RW, Haasen P. Physical metallurgy. 4th ed. Cambridge: Univ Press; 1996

[5] Sandler S. I., *Chemical and Engineering Thermodynamics*. 1999 John Wiley & Sons, Inc., p 557.

[6] Koningsveld K., Stockmayer W. H.,Nies, E., *Polymer Phase Diagrams: A Textbook*. 2001 Oxford, pp 46-47.

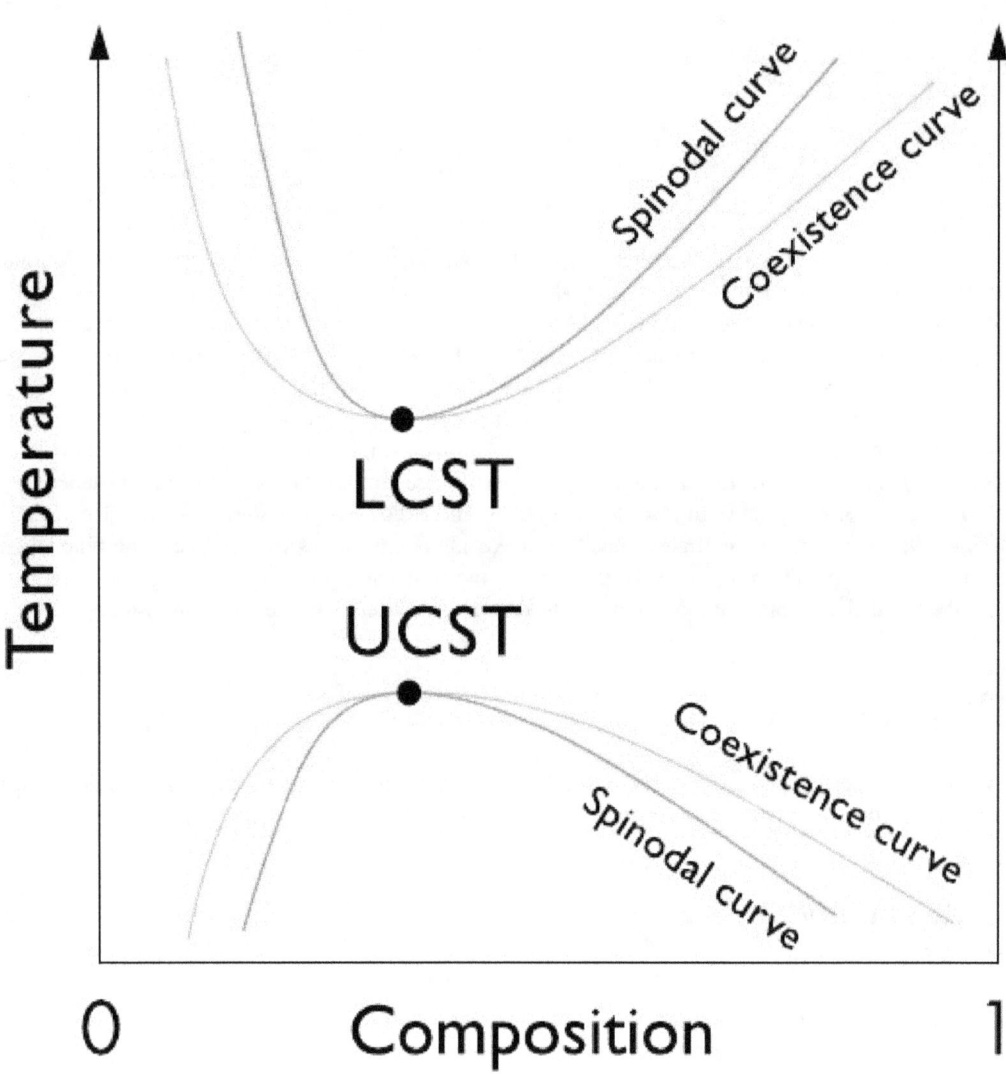

A phase diagram displaying spinodal curves, within the coexistence curves and two critical points: an upper and lower critical solution temperature.

Chapter 18

Superconductivity

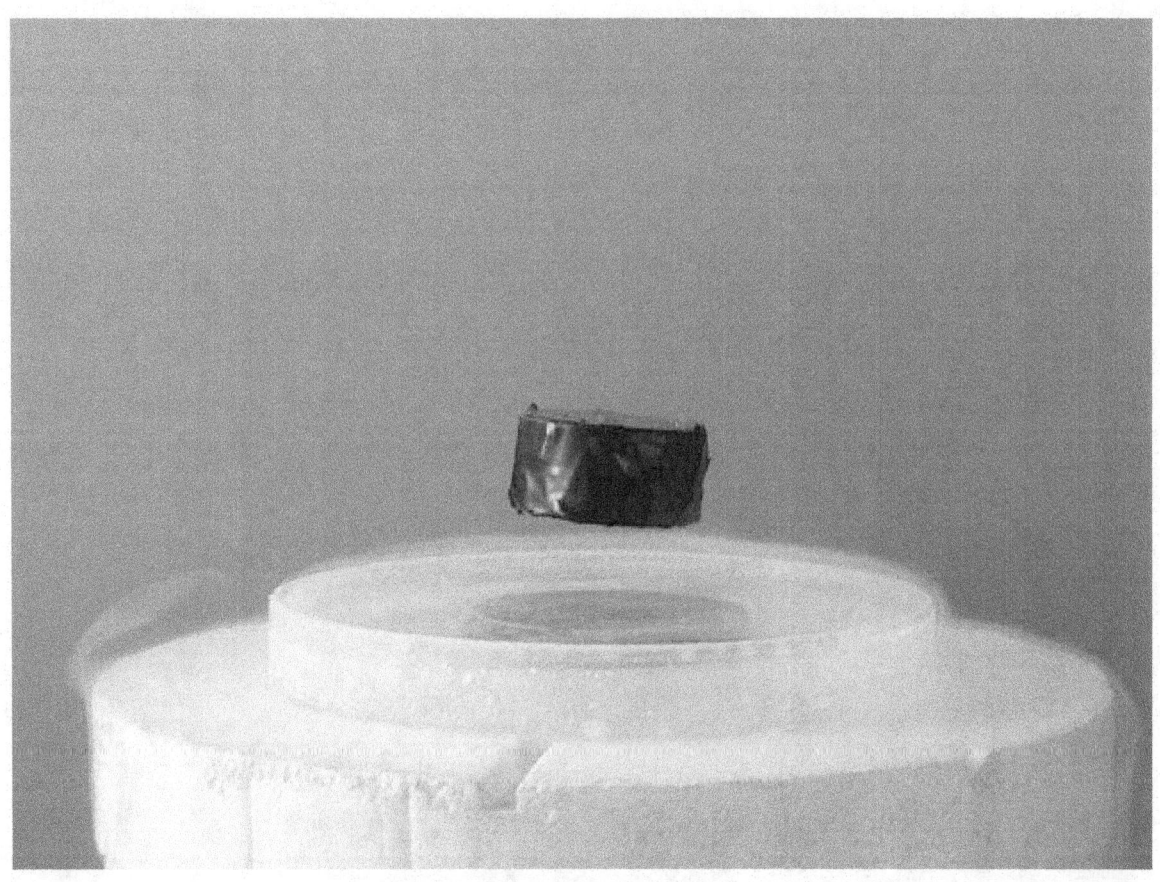

A magnet levitating above a high-temperature superconductor, cooled with liquid nitrogen. Persistent electric current flows on the surface of the superconductor, acting to exclude the magnetic field of the magnet (Faraday's law of induction). This current effectively forms an electromagnet that repels the magnet.

Superconductivity is a phenomenon of exactly zero electrical resistance and expulsion of magnetic fields occurring in certain materials when cooled below a characteristic critical temperature. It was discovered by Dutch physicist Heike Kamerlingh Onnes on April 8, 1911 in Leiden. Like ferromagnetism and atomic spectral lines, superconductivity is a quantum mechanical phenomenon. It is characterized by the Meissner effect, the complete ejection of magnetic field lines from the interior of the superconductor as it transitions into the superconducting state. The occurrence of the Meissner effect indicates that superconductivity cannot be understood simply as the idealization of *perfect conductivity* in classical

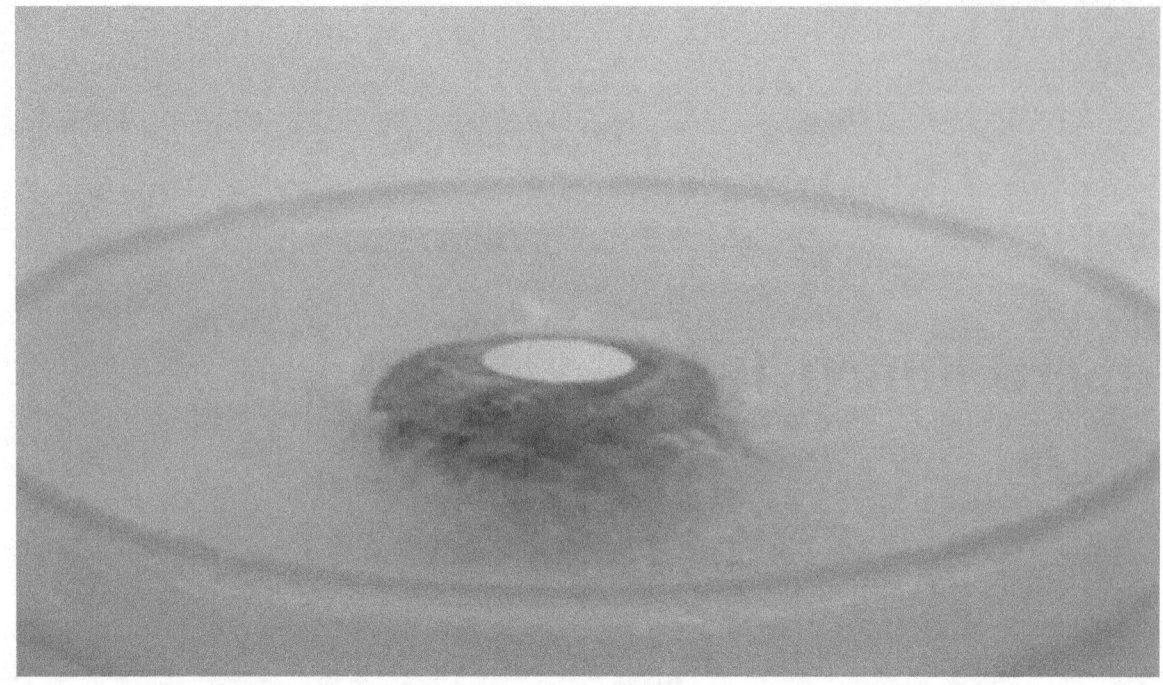

Video of a Meissner effect in a high temperature superconductor (black pellet) with a NdFeB magnet (metallic)

A high-temperature superconductor levitating above a magnet

physics.

The electrical resistivity of a metallic conductor decreases gradually as temperature is lowered. In ordinary conductors, such as copper or silver, this decrease is limited by impurities and other defects. Even near absolute zero, a real sample of a normal conductor shows some resistance. In a superconductor, the resistance drops abruptly to zero when the material is cooled below its critical temperature. An electric current flowing through a loop of superconducting wire can persist indefinitely with no power source.[1][2][3][4]

In 1986, it was discovered that some cuprate-perovskite ceramic materials have a critical temperature above 90 K (−183 °C).[5] Such a high transition temperature is theoretically impossible for a conventional superconductor, leading the materials to be termed high-temperature superconductors. Liquid nitrogen boils at 77 K, and superconduction at higher temperatures than this facilitates many experiments and applications that are less practical at lower temperatures.

18.1 Classification

Main article: Superconductor classification

There are many criteria by which superconductors are classified. The most common are:

- **Response to a magnetic field**: A superconductor can be *Type I*, meaning it has a single critical field, above which all superconductivity is lost; or *Type II*, meaning it has two critical fields, between which it allows partial penetration of the magnetic field.

- **By theory of operation**: It is *conventional* if it can be explained by the BCS theory or its derivatives, or *unconventional*, otherwise.

- **By critical temperature**: A superconductor is generally considered *high temperature* if it reaches a superconducting state when cooled using liquid nitrogen – that is, at only $Tc > 77$ K) – or *low temperature* if more aggressive cooling techniques are required to reach its critical temperature.

- **By material**: Superconductor material classes include chemical elements (e.g. mercury or lead), alloys (such as niobium-titanium, germanium-niobium, and niobium nitride), ceramics (YBCO and magnesium diboride), or organic superconductors (fullerenes and carbon nanotubes; though perhaps these examples should be included among the chemical elements, as they are composed entirely of carbon).

18.2 Elementary properties of superconductors

Most of the physical properties of superconductors vary from material to material, such as the heat capacity and the critical temperature, critical field, and critical current density at which superconductivity is destroyed.

On the other hand, there is a class of properties that are independent of the underlying material. For instance, all superconductors have *exactly* zero resistivity to low applied currents when there is no magnetic field present or if the applied field does not exceed a critical value. The existence of these "universal" properties implies that superconductivity is a thermodynamic phase, and thus possesses certain distinguishing properties which are largely independent of microscopic details.

18.2.1 Zero electrical DC resistance

The simplest method to measure the electrical resistance of a sample of some material is to place it in an electrical circuit in series with a current source I and measure the resulting voltage V across the sample. The resistance of the sample is given by Ohm's law as $R = V/I$. If the voltage is zero, this means that the resistance is zero.

Superconductors are also able to maintain a current with no applied voltage whatsoever, a property exploited in superconducting electromagnets such as those found in MRI machines. Experiments have demonstrated that currents in superconducting

Electric cables for accelerators at CERN. Both the massive and slim cables are rated for 12,500 A. Top: *conventional cables for LEP;* bottom: *superconductor-based cables for the LHC*

coils can persist for years without any measurable degradation. Experimental evidence points to a current lifetime of at least 100,000 years. Theoretical estimates for the lifetime of a persistent current can exceed the estimated lifetime of the universe, depending on the wire geometry and the temperature.[3]

In a normal conductor, an electric current may be visualized as a fluid of electrons moving across a heavy ionic lattice. The electrons are constantly colliding with the ions in the lattice, and during each collision some of the energy carried by the current is absorbed by the lattice and converted into heat, which is essentially the vibrational kinetic energy of the lattice ions. As a result, the energy carried by the current is constantly being dissipated. This is the phenomenon of electrical resistance and Joule heating.

The situation is different in a superconductor. In a conventional superconductor, the electronic fluid cannot be resolved into individual electrons. Instead, it consists of bound *pairs* of electrons known as Cooper pairs. This pairing is caused by an attractive force between electrons from the exchange of phonons. Due to quantum mechanics, the energy spectrum of this Cooper pair fluid possesses an *energy gap*, meaning there is a minimum amount of energy ΔE that must be supplied in order to excite the fluid. Therefore, if ΔE is larger than the thermal energy of the lattice, given by kT, where k is Boltzmann's constant and T is the temperature, the fluid will not be scattered by the lattice. The Cooper pair fluid is thus a superfluid, meaning it can flow without energy dissipation.

In a class of superconductors known as type II superconductors, including all known high-temperature superconductors, an extremely small amount of resistivity appears at temperatures not too far below the nominal superconducting transition when an electric current is applied in conjunction with a strong magnetic field, which may be caused by the electric current. This is due to the motion of magnetic vortices in the electronic superfluid, which dissipates some of the energy carried by the current. If the current is sufficiently small, the vortices are stationary, and the resistivity vanishes. The resistance due to this effect is tiny compared with that of non-superconducting materials, but must be taken into account in sensitive experiments. However, as the temperature decreases far enough below the nominal superconducting transition, these vortices can become frozen into a disordered but stationary phase known as a "vortex glass". Below this vortex glass transition temperature, the resistance of the material becomes truly zero.

18.2.2 Superconducting phase transition

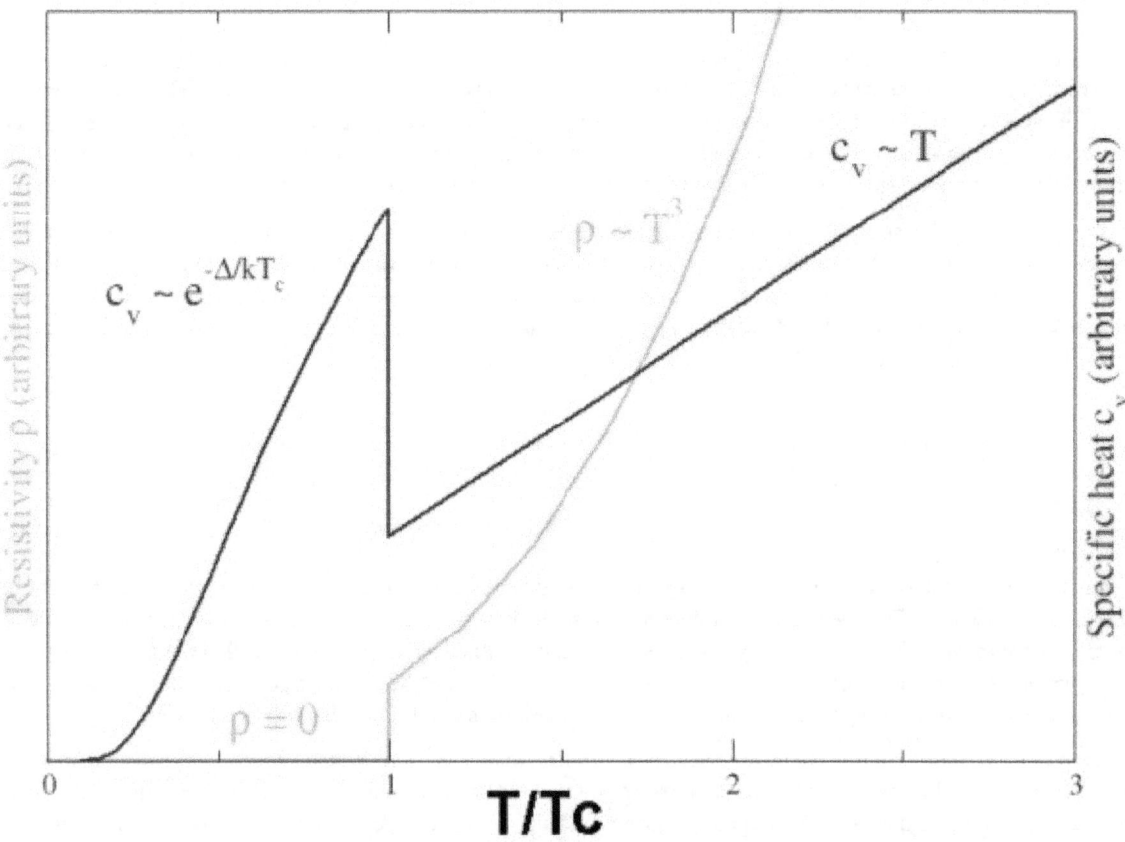

Behavior of heat capacity (cv, blue) and resistivity (ρ, green) at the superconducting phase transition

In superconducting materials, the characteristics of superconductivity appear when the temperature T is lowered below a critical temperature T_c. The value of this critical temperature varies from material to material. Conventional superconductors usually have critical temperatures ranging from around 20 K to less than 1 K. Solid mercury, for example, has a critical temperature of 4.2 K. As of 2009, the highest critical temperature found for a conventional superconductor is 39 K for magnesium diboride (MgB_2),[6][7] although this material displays enough exotic properties that there is some doubt about classifying it as a "conventional" superconductor.[8] Cuprate superconductors can have much higher critical temperatures: $YBa_2Cu_3O_7$, one of the first cuprate superconductors to be discovered, has a critical temperature of 92 K, and mercury-based cuprates have been found with critical temperatures in excess of 130 K. The explanation for these high critical temperatures remains unknown. Electron pairing due to phonon exchanges explains superconductivity in conventional superconductors, but it does not explain superconductivity in the newer superconductors that have a very high critical temperature.

Similarly, at a fixed temperature below the critical temperature, superconducting materials cease to superconduct when an external magnetic field is applied which is greater than the *critical magnetic field*. This is because the Gibbs free energy of the superconducting phase increases quadratically with the magnetic field while the free energy of the normal phase is roughly independent of the magnetic field. If the material superconducts in the absence of a field, then the superconducting phase free energy is lower than that of the normal phase and so for some finite value of the magnetic field (proportional to the square root of the difference of the free energies at zero magnetic field) the two free energies will be equal and a phase transition to the normal phase will occur. More generally, a higher temperature and a stronger magnetic field lead to a smaller fraction of the electrons in the superconducting band and consequently a longer London penetration depth of external magnetic fields and currents. The penetration depth becomes infinite at the phase transition.

The onset of superconductivity is accompanied by abrupt changes in various physical properties, which is the hallmark of a phase transition. For example, the electronic heat capacity is proportional to the temperature in the normal (non-superconducting) regime. At the superconducting transition, it suffers a discontinuous jump and thereafter ceases to be linear. At low temperatures, it varies instead as $e^{-\alpha/T}$ for some constant, α. This exponential behavior is one of the pieces of evidence for the existence of the energy gap.

The order of the superconducting phase transition was long a matter of debate. Experiments indicate that the transition is second-order, meaning there is no latent heat. However, in the presence of an external magnetic field there is latent heat, because the superconducting phase has a lower entropy below the critical temperature than the normal phase. It has been experimentally demonstrated[9] that, as a consequence, when the magnetic field is increased beyond the critical field, the resulting phase transition leads to a decrease in the temperature of the superconducting material.

Calculations in the 1970s suggested that it may actually be weakly first-order due to the effect of long-range fluctuations in the electromagnetic field. In the 1980s it was shown theoretically with the help of a disorder field theory, in which the vortex lines of the superconductor play a major role, that the transition is of second order within the type II regime and of first order (i.e., latent heat) within the type I regime, and that the two regions are separated by a tricritical point.[10] The results were strongly supported by Monte Carlo computer simulations.[11]

18.2.3   Meissner effect

Main article: Meissner effect

When a superconductor is placed in a weak external magnetic field **H**, and cooled below its transition temperature, the magnetic field is ejected. The Meissner effect does not cause the field to be completely ejected but instead the field penetrates the superconductor but only to a very small distance, characterized by a parameter λ, called the London penetration depth, decaying exponentially to zero within the bulk of the material. The Meissner effect is a defining characteristic of superconductivity. For most superconductors, the London penetration depth is on the order of 100 nm.

The Meissner effect is sometimes confused with the kind of diamagnetism one would expect in a perfect electrical conductor: according to Lenz's law, when a *changing* magnetic field is applied to a conductor, it will induce an electric current in the conductor that creates an opposing magnetic field. In a perfect conductor, an arbitrarily large current can be induced, and the resulting magnetic field exactly cancels the applied field.

The Meissner effect is distinct from this—it is the spontaneous expulsion which occurs during transition to superconductivity. Suppose we have a material in its normal state, containing a constant internal magnetic field. When the material is cooled below the critical temperature, we would observe the abrupt expulsion of the internal magnetic field, which we would not expect based on Lenz's law.

The Meissner effect was given a phenomenological explanation by the brothers Fritz and Heinz London, who showed that the electromagnetic free energy in a superconductor is minimized provided

$$\nabla^2 \mathbf{H} = \lambda^{-2} \mathbf{H}$$

where **H** is the magnetic field and λ is the London penetration depth.

This equation, which is known as the London equation, predicts that the magnetic field in a superconductor decays exponentially from whatever value it possesses at the surface.

A superconductor with little or no magnetic field within it is said to be in the Meissner state. The Meissner state breaks down when the applied magnetic field is too large. Superconductors can be divided into two classes according to how this breakdown occurs. In Type I superconductors, superconductivity is abruptly destroyed when the strength of the applied field rises above a critical value Hc. Depending on the geometry of the sample, one may obtain an intermediate state[12] consisting of a baroque pattern[13] of regions of normal material carrying a magnetic field mixed with regions of superconducting material containing no field. In Type II superconductors, raising the applied field past a critical value Hc_1 leads to a mixed state (also known as the vortex state) in which an increasing amount of magnetic flux penetrates the material, but there remains no resistance to the flow of electric current as long as the current is not too large. At a second critical field strength Hc_2, superconductivity is destroyed. The mixed state is actually caused by vortices in the electronic superfluid, sometimes called fluxons because the flux carried by these vortices is quantized. Most pure elemental superconductors, except niobium and carbon nanotubes, are Type I, while almost all impure and compound superconductors are Type II.

18.2.4   London moment

Conversely, a spinning superconductor generates a magnetic field, precisely aligned with the spin axis. The effect, the London moment, was put to good use in Gravity Probe B. This experiment measured the magnetic fields of four superconducting gyroscopes to determine their spin axes. This was critical to the experiment since it is one of the few ways to accurately determine the spin axis of an otherwise featureless sphere.

18.3   History of superconductivity

Heike Kamerlingh Onnes (right), the discoverer of superconductivity. Paul Ehrenfest, Hendrik Lorentz, Niels Bohr stand to his left.

Main article: History of superconductivity

Superconductivity was discovered on April 8, 1911 by Heike Kamerlingh Onnes, who was studying the resistance of solid mercury at cryogenic temperatures using the recently produced liquid helium as a refrigerant. At the temperature of

4.2 K, he observed that the resistance abruptly disappeared.[14] In the same experiment, he also observed the superfluid transition of helium at 2.2 K, without recognizing its significance. The precise date and circumstances of the discovery were only reconstructed a century later, when Onnes's notebook was found.[15] In subsequent decades, superconductivity was observed in several other materials. In 1913, lead was found to superconduct at 7 K, and in 1941 niobium nitride was found to superconduct at 16 K.

Great efforts have been devoted to finding out how and why superconductivity works; the important step occurred in 1933, when Meissner and Ochsenfeld discovered that superconductors expelled applied magnetic fields, a phenomenon which has come to be known as the Meissner effect.[16] In 1935, Fritz and Heinz London showed that the Meissner effect was a consequence of the minimization of the electromagnetic free energy carried by superconducting current.[17]

18.3.1   London theory

The first phenomenological theory of superconductivity was London theory. It was put forward by the brothers Fritz and Heinz London in 1935, shortly after the discovery that magnetic fields are expelled from superconductors. A major triumph of the equations of this theory is their ability to explain the Meissner effect,[18] wherein a material exponentially expels all internal magnetic fields as it crosses the superconducting threshold. By using the London equation, one can obtain the dependence of the magnetic field inside the superconductor on the distance to the surface.[19]

There are two London equations:

$$\frac{\partial \mathbf{j}_s}{\partial t} = \frac{n_s e^2}{m}\mathbf{E}, \qquad \nabla \times \mathbf{j}_s = -\frac{n_s e^2}{m}\mathbf{B}.$$

The first equation follows from Newton's second law for superconducting electrons.

18.3.2   Conventional theories (1950s)

During the 1950s, theoretical condensed matter physicists arrived at a solid understanding of "conventional" superconductivity, through a pair of remarkable and important theories: the phenomenological Ginzburg-Landau theory (1950) and the microscopic BCS theory (1957).[20][21]

In 1950, the phenomenological Ginzburg-Landau theory of superconductivity was devised by Landau and Ginzburg.[22] This theory, which combined Landau's theory of second-order phase transitions with a Schrödinger-like wave equation, had great success in explaining the macroscopic properties of superconductors. In particular, Abrikosov showed that Ginzburg-Landau theory predicts the division of superconductors into the two categories now referred to as Type I and Type II. Abrikosov and Ginzburg were awarded the 2003 Nobel Prize for their work (Landau had received the 1962 Nobel Prize for other work, and died in 1968). The four-dimensional extension of the Ginzburg-Landau theory, the Coleman-Weinberg model, is important in quantum field theory and cosmology.

Also in 1950, Maxwell and Reynolds *et al.* found that the critical temperature of a superconductor depends on the isotopic mass of the constituent element.[23][24] This important discovery pointed to the electron-phonon interaction as the microscopic mechanism responsible for superconductivity.

The complete microscopic theory of superconductivity was finally proposed in 1957 by Bardeen, Cooper and Schrieffer.[21] This BCS theory explained the superconducting current as a superfluid of Cooper pairs, pairs of electrons interacting through the exchange of phonons. For this work, the authors were awarded the Nobel Prize in 1972.

The BCS theory was set on a firmer footing in 1958, when N. N. Bogolyubov showed that the BCS wavefunction, which had originally been derived from a variational argument, could be obtained using a canonical transformation of the electronic Hamiltonian.[25] In 1959, Lev Gor'kov showed that the BCS theory reduced to the Ginzburg-Landau theory close to the critical temperature.[26][27]

Generalizations of BCS theory for conventional superconductors form the basis for understanding of the phenomenon of superfluidity, because they fall into the lambda transition universality class. The extent to which such generalizations can be applied to unconventional superconductors is still controversial.

18.3.3 Further history

The first practical application of superconductivity was developed in 1954 with Dudley Allen Buck's invention of the cryotron.[28] Two superconductors with greatly different values of critical magnetic field are combined to produce a fast, simple, switch for computer elements.

Soon after discovering superconductivity in 1911, Kamerlingh Onnes attempted to make an electromagnet with superconducting windings but found that relatively low magnetic fields destroyed superconductivity in the materials he investigated. Much later, in 1955, G.B. Yntema [29] succeeded in constructing a small 0.7-tesla iron-core electromagnet with superconducting niobium wire windings. Then, in 1961, J.E. Kunzler, E. Buehler, F.S.L. Hsu, and J.H. Wernick [30] made the startling discovery that, at 4.2 degrees kelvin, a compound consisting of three parts niobium and one part tin, was capable of supporting a current density of more than 100,000 amperes per square centimeter in a magnetic field of 8.8 tesla. Despite being brittle and difficult to fabricate, niobium-tin has since proved extremely useful in supermagnets generating magnetic fields as high as 20 tesla. In 1962 T.G. Berlincourt and R.R. Hake [31][32] discovered that alloys of niobium and titanium are suitable for applications up to 10 tesla. Promptly thereafter, commercial production of niobium-titanium supermagnet wire commenced at Westinghouse Electric Corporation and at Wah Chang Corporation. Although niobium-titanium boasts less-impressive superconducting properties than those of niobium-tin, niobium-titanium has, nevertheless, become the most widely-used "workhorse" supermagnet material, in large measure a consequence of its very-high ductility and ease of fabrication. However, both niobium-tin and niobium-titanium find wide application in MRI medical imagers, bending and focusing magnets for enormous high-energy-particle accelerators, and a host of other applications. Conectus, a European superconductivity consortium, estimated that in 2014, global economic activity for which superconductivity was indispensable amounted to about five billion euros, with MRI systems accounting for about 80% of that total.

In 1962, Josephson made the important theoretical prediction that a supercurrent can flow between two pieces of superconductor separated by a thin layer of insulator.[33] This phenomenon, now called the Josephson effect, is exploited by superconducting devices such as SQUIDs. It is used in the most accurate available measurements of the magnetic flux quantum $\Phi_0 = h/(2e)$, where h is the Planck constant. Coupled with the quantum Hall resistivity, this leads to a precise measurement of the Planck constant. Josephson was awarded the Nobel Prize for this work in 1973.

In 2008, it was proposed that the same mechanism that produces superconductivity could produce a superinsulator state in some materials, with almost infinite electrical resistance.[34]

18.4 High-temperature superconductivity

Main article: High-temperature superconductivity

Until 1986, physicists had believed that BCS theory forbade superconductivity at temperatures above about 30 K. In that year, Bednorz and Müller discovered superconductivity in a lanthanum-based cuprate perovskite material, which had a transition temperature of 35 K (Nobel Prize in Physics, 1987).[5] It was soon found that replacing the lanthanum with yttrium (i.e., making YBCO) raised the critical temperature to 92 K.[35]

This temperature jump is particularly significant, since it allows liquid nitrogen as a refrigerant, replacing liquid helium.[35] This can be important commercially because liquid nitrogen can be produced relatively cheaply, even on-site. Also, the higher temperatures help avoid some of the problems that arise at liquid helium temperatures, such as the formation of plugs of frozen air that can block cryogenic lines and cause unanticipated and potentially hazardous pressure buildup.[36][37]

Many other cuprate superconductors have since been discovered, and the theory of superconductivity in these materials is one of the major outstanding challenges of theoretical condensed matter physics.[38] There are currently two main hypotheses – the resonating-valence-bond theory, and spin fluctuation which has the most support in the research community.[39] The second hypothesis proposed that electron pairing in high-temperature superconductors is mediated by short-range spin waves known as paramagnons.[40][41]

Since about 1993, the highest temperature superconductor was a ceramic material consisting of mercury, barium, calcium, copper and oxygen ($HgBa_2Ca_2Cu_3O_{8+}\delta$) with T_c = 133–138 K.[42][43] The latter experiment (138 K) still awaits experimental confirmation, however.

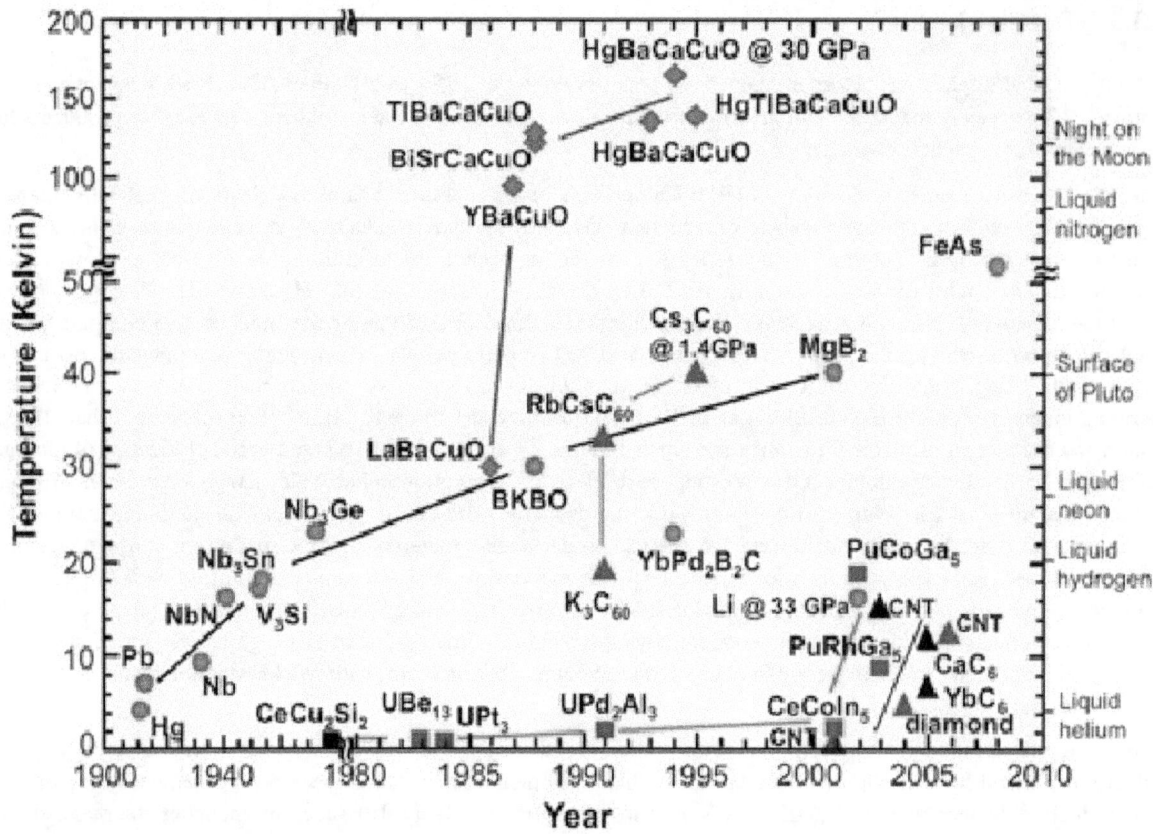

Timeline of superconducting materials

In February 2008, an iron-based family of high-temperature superconductors was discovered.[44][45] Hideo Hosono, of the Tokyo Institute of Technology, and colleagues found lanthanum oxygen fluorine iron arsenide ($LaO_{1-x}F_xFeAs$), an oxypnictide that superconducts below 26 K. Replacing the lanthanum in $LaO_{1-x}F_xFeAs$ with samarium leads to superconductors that work at 55 K.[46]

In May 2014, hydrogen sulfide (H
2S) was predicted to be a high-temperature superconductor with a transition temperate of 80 at 160 gigapascals.[47] In 2015, H
2S has been observed to exhibit superconductivity at below 203 K but at extremely high pressures — around 150 gigapascals.[48]

18.5   Applications

Main article: Technological applications of superconductivity
 Superconducting magnets are some of the most powerful electromagnets known. They are used in MRI/NMR machines, mass spectrometers, and the beam-steering magnets used in particle accelerators. They can also be used for magnetic separation, where weakly magnetic particles are extracted from a background of less or non-magnetic particles, as in the pigment industries.

In the 1950s and 1960s, superconductors were used to build experimental digital computers using cryotron switches. More recently, superconductors have been used to make digital circuits based on rapid single flux quantum technology and RF and microwave filters for mobile phone base stations.

Superconductors are used to build Josephson junctions which are the building blocks of SQUIDs (superconducting quantum interference devices), the most sensitive magnetometers known. SQUIDs are used in scanning SQUID microscopes

Video of superconducting levitation of YBCO

and magnetoencephalography. Series of Josephson devices are used to realize the SI volt. Depending on the particular mode of operation, a superconductor-insulator-superconductor Josephson junction can be used as a photon detector or as a mixer. The large resistance change at the transition from the normal- to the superconducting state is used to build thermometers in cryogenic micro-calorimeter photon detectors. The same effect is used in ultrasensitive bolometers made from superconducting materials.

Other early markets are arising where the relative efficiency, size and weight advantages of devices based on high-temperature superconductivity outweigh the additional costs involved. For example in wind turbines the lower weight and volume of superconducting generators could lead to savings in construction and tower costs, offsetting the higher costs for the generator and lowering the total LCOE.[49]

Promising future applications include high-performance smart grid, electric power transmission, transformers, power storage devices, electric motors (e.g. for vehicle propulsion, as in vactrains or maglev trains), magnetic levitation devices, fault current limiters, enhancing spintronic devices with superconducting materials,[50] and superconducting magnetic refrigeration. However, superconductivity is sensitive to moving magnetic fields so applications that use alternating current (e.g. transformers) will be more difficult to develop than those that rely upon direct current. Compared to to traditional power lines superconducting transmission lines are more efficient and require only a fraction of the space, which would not only lead to a better environmental performance but could also improve public acceptance for expansion of the electric grid.[51]

18.6 Nobel Prizes for superconductivity

- Heike Kamerlingh Onnes (1913), "for his investigations on the properties of matter at low temperatures which led, inter alia, to the production of liquid helium"

- John Bardeen, Leon N. Cooper, and J. Robert Schrieffer (1972), "for their jointly developed theory of superconductivity, usually called the BCS-theory"

- Leo Esaki, Ivar Giaever, and Brian D. Josephson (1973), "for their experimental discoveries regarding tunneling phenomena in semiconductors and superconductors, respectively," and "for his theoretical predictions of the properties of a supercurrent through a tunnel barrier, in particular those phenomena which are generally known as the Josephson effects"

- Georg Bednorz and K. Alex Müller (1987), "for their important break-through in the discovery of superconductivity in ceramic materials"

- Alexei A. Abrikosov, Vitaly L. Ginzburg, and Anthony J. Leggett (2003), "for pioneering contributions to the theory of superconductors and superfluids"[52]

18.7   See also

- Andreev reflection

- Charge transfer complex

- Color superconductivity in quarks

- Composite Reaction Texturing

- Conventional superconductor

- Covalent superconductors

- Flux pumping

- High-temperature superconductivity

- Homes's law

- Iron-based superconductor

- Kondo effect

- List of superconductors

- Little-Parks effect

- Magnetic levitation

- Macroscopic quantum phenomena

- Magnetic sail

- National Superconducting Cyclotron Laboratory

- Oxypnictide

- Persistent current

- Proximity effect

- Room-temperature superconductor

- Rutherford cable

- Spallation Neutron Source

- Superconducting RF

- Superconductor classification

- Superfluid film

- Superfluidity

- Superstripes

- Technological applications of superconductivity

- Timeline of low-temperature technology

- Type-I superconductor

- Type-II superconductor

- Unconventional superconductor

- BCS theory

- Bean's critical state model

18.8   References

[1] John Bardeen; Leon Cooper; J. R. Schriffer (December 1, 1957). "Theory of Superconductivity". *Physical Review* 8 (5): 1178. Bibcode:1957PhRv..108.1175B. doi:10.1103/physrev.108.1175. ISBN 9780677000800. Retrieved June 6, 2014. reprinted in Nikolaĭ Nikolaevich Bogoliubov (1963) *The Theory of Superconductivity, Vol. 4*, CRC Press, ISBN 0677000804, p. 73

[2] John Daintith (2009). *The Facts on File Dictionary of Physics* (4th ed.). Infobase Publishing. p. 238. ISBN 1438109490.

[3] John C. Gallop (1990). *SQUIDS, the Josephson Effects and Superconducting Electronics*. CRC Press. pp. 3, 20. ISBN 0-7503-0051-5.

[4] Durrant, Alan (2000). *Quantum Physics of Matter*. CRC Press. pp. 102–103. ISBN 0750307218.

[5] J. G. Bednorz & K. A. Müller (1986). "Possible high T_c superconductivity in the Ba–La–Cu–O system". *Z. Physik, B* 64 (1): 189–193. Bibcode:1986ZPhyB..64..189B. doi:10.1007/BF01303701.

[6] Jun Nagamatsu; Norimasa Nakagawa; Takahiro Muranaka; Yuji Zenitani; et al. (2001). "Superconductivity at 39 K in magnesium diboride". *Nature* 410 (6824): 63–4. Bibcode:2001Natur.410...63N. doi:10.1038/35065039. PMID 11242039.

[7] Paul Preuss (14 August 2002). "A most unusual superconductor and how it works: first-principles calculation explains the strange behavior of magnesium diboride". *Research News* (Lawrence Berkeley National Laboratory). Retrieved 2009-10-28.

[8] Hamish Johnston (17 February 2009). "Type-1.5 superconductor shows its stripes". *Physics World* (Institute of Physics). Retrieved 2009-10-28.

[9] R. L. Dolecek (1954). "Adiabatic Magnetization of a Superconducting Sphere". *Physical Review* 96 (1): 25–28. doi:10.1103/PhysRev.96.25.

[10] H. Kleinert (1982). "Disorder Version of the Abelian Higgs Model and the Order of the Superconductive Phase Transition" (PDF). *Lettere al Nuovo Cimento* 35 (13): 405–412. doi:10.1007/BF02754760.

[11] J. Hove; S. Mo; A. Sudbo (2002). "Vortex interactions and thermally induced crossover from type-I to type-II superconductivity" (PDF). *Physical Review B* 66 (6): 064524. arXiv:cond-mat/0202215. Bibcode:2002PhRvB..66f4524H. doi:10.1103/

[12] Lev D. Landau; Evgeny M. Lifschitz (1984). *Electrodynamics of Continuous Media*. Course of Theoretical Physics 8. Oxford: Butterworth-Heinemann. ISBN 0-7506-2634-8.

[13] David J. E. Callaway (1990). "On the remarkable structure of the superconducting intermediate state". *Nuclear Physics B* 344 (3): 627–645. Bibcode:1990NuPhB.344..627C. doi:10.1016/0550-3213(90)90672-Z.

[14] H. K. Onnes (1911). "The resistance of pure mercury at helium temperatures". *Commun. Phys. Lab. Univ. Leiden* **12**: 120.

[15] Dirk vanDelft & Peter Kes (September 2010). "The Discovery of Superconductivity" (PDF). *Physics Today* (American Institute of Physics).

[16] W. Meissner & R. Ochsenfeld (1933). "Ein neuer Effekt bei Eintritt der Supraleitfähigkeit". *Naturwissenschaften* **21** (44): 787–788. Bibcode:1933NW.....21..787M. doi:10.1007/BF01504252.

[17] F. London & H. London (1935). "The Electromagnetic Equations of the Supraconductor". *Proceedings of the Royal Society of London A* **149** (866): 71–88. Bibcode:1935RSPSA.149...71L. doi:10.1098/rspa.1935.0048. JSTOR 96265.

[18] Meissner, W.; R. Ochsenfeld (1933). "Ein neuer Effekt bei Eintritt der Supraleitfähigkeit". *Naturwissenschaften* **21** (44): 787–788. Bibcode:1933NW.....21..787M. doi:10.1007/BF01504252.

[19] "The London equations". The Open University. Retrieved 2011-10-16.

[20] J. Bardeen; L. N. Cooper & J. R. Schrieffer (1957). "Microscopic Theory of Superconductivity". *Physical Review* **106** (1): 162–164. Bibcode:1957PhRv..106..162B. doi:10.1103/PhysRev.106.162.

[21] J. Bardeen; L. N. Cooper & J. R. Schrieffer (1957). "Theory of Superconductivity". *Physical Review* **108** (5): 1175–1205. Bibcode:1957PhRv..108.1175B. doi:10.1103/PhysRev.108.1175.

[22] V. L. Ginzburg & L.D. Landau (1950). "On the theory of superconductivity". *Zhurnal Eksperimental'noi i Teoreticheskoi Fiziki* **20**: 1064.

[23] E. Maxwell (1950). "Isotope Eff ect in the Superconductivity of Mercury". *Physical Review* **78** (4): 477. Bibcode: doi:10.1103/PhysRev.78.477.

[24] C. A. Reynolds; B. Serin; W. H. Wright & L. B. Nesbitt (1950). "Superconductivity of Isotopes of Mercury". *Physical Review* **78** (4): 487. Bibcode:1950PhRv...78..487R. doi:10.1103/PhysRev.78.487.

[25] N. N. Bogoliubov (1958). "A new method in the theory of superconductivity". *Zhurnal Eksperimental'noi i Teoreticheskoi Fiziki* **34**: 58.

[26] L. P. Gor'kov (1959). "Microscopic derivation of the Ginzburg—Landau equations in the theory of superconductivity". *Zhurnal Eksperimental'noi i Teoreticheskoi Fiziki* **36**: 1364.

[27] M. Combescot; W.V. Pogosov and O. Betbeder-Matibet (2013). "BCS ansatz for superconductivity in the light of the Bogoliubov approach and the Richardson–Gaudin exact wave function". *Physica C: Superconductivity* **485**: 47–57. arXiv:1111.4781. Bibcode:2013PhyC..485...47C. doi:10.1016/j.physc.2012.10.011. Retrieved 11 August 2014.

[28] Buck, Dudley A. "The Cryotron - A Superconductive Computer Component" (PDF). Lincoln Laboratory, Massachusetts Institute of Technology. Retrieved 10 August 2014.

[29] G.B.Yntema (1955). "Superconducting Winding for Electromagnet". *Physical Review* **98** (4): 1197. Bibcode:1955 doi:10.1103/PhysRev.98.1144.

[30] J.E. Kunzler, E. Buehler, F.L.S. Hsu, and J.H. Wernick (1961). "Superconductivity in Nb3Sn at High Current Density in a Magnetic Field of 88 kgauss". *Physical Review Letters* **6** (3): 89–91. Bibcode:1961PhRvL...6...89K. doi:10.1103/PhysRevLett.6.89. line feed character in |title= at position 65 (help)

[31] T.G. Berlincourt and R.R. Hake (1962). "Pulsed-Magnetic-Field Studies of Superconducting Transition Metal Alloys at High and Low Current Densities". *Bulletin of the American Physical Society* **II–7**: 408. line feed character in |title= at position 60 (help)

[32] T.G. Berlincourt (1987). "Emergence of Nb-Ti as Supermagnet Material". *Cryogenics* **27** (6): 283–289. Bibcode doi:10.1016/0011-2275(87)90057-9.

[33] B. D. Josephson (1962). "Possible new eff ects in superconductive tunnelling". *Physics Letters* **1** (7): 251– doi:10.1016/0031-9163(62)91369-0.

[34] "Newly discovered fundamental state of matter, a superinsulator, has been created.". Science Daily. April 9, 2008. Retrieved 2008-10-23.

[35] M. K. Wu; et al. (1987). "Superconductivity at 93 K in a New Mixed-Phase Y-Ba-Cu-O Compound System at Ambient Pressure". *Physical Review Letters* **58** (9): 908–910. Bibcode:1987PhRvL..58..908W. doi:10.1103/PhysRevLett.58.908. PMID 10035069.

[36] "Introduction to Liquid Helium". *"Cryogenics and Fluid Branch"*. Goddard Space Flight Center, NASA.

[37] "Section 4.1 "Air plug in the fill line"". *"Superconducting Rock Magnetometer Cryogenic System Manual"*. 2G Enterprises. Archived from the original on May 6, 2009. Retrieved 9 October 2012.

[38] Alexei A. Abrikosov (8 December 2003). "type II Superconductors and the Vortex Lattice". *Nobel Lecture*.

[39] Adam Mann (Jul 20, 2011). "High-temperature superconductivity at 25: Still in suspense". *Nature* **475** (7356): 280–2. Bibcode:2011Natur.475..280M. doi:10.1038/475280a. PMID 21776057.

[40] Pines, D. (2002), "The Spin Fluctuation Model for High Temperature Superconductivity: Progress and Prospects", *The Gap Symmetry and Fluctuations in High-Tc Superconductors*, NATO Science Series: B: **371**, New York: Kluwer Academic, pp. 111–142, doi:10.1007/0-306-47081-0_7, ISBN 0-306-45934-5

[41] P. Monthoux; A. V. Balatsky & D. Pines (1991). "Toward a theory of high-temperature superconductivity in the antiferromagnetically correlated cuprate oxides". *Phys. Rev. Lett.* **67** (24): 3448–3451. Bibcode:1991PhRvL..67.3448M. doi:10.1103/PhysRevLett.67.3448.PMID 10044736.

[42] A. Schilling; et al. (1993). "Superconductivity above 130 K in the Hg–Ba–Ca–Cu–O system". *Nature* **363** (6424): 56–58. Bibcode:1993Natur.363..56C. doi:10.1038/363056a0.

[43] P. Dai; B. C. Chakoumakos; G. F. Sun; K. W. Wong; et al. (1995). "Synthesis and neutron powder diffraction study of the superconductor $HgBa_2Ca_2Cu_3O_{8+\delta}$ by Tl substitution". *Physica C* **243** (3–4): 201–206. Bibcode:1995PhyC..243..201D. doi:10.1016/0921-4534(94)02461-8.

[44] Hiroki Takahashi; Kazumi Igawa; Kazunobu Arii; Yoichi Kamihara; et al. (2008). "Superconductivity at 43 K in an iron-based layered compound $LaO_{1-x}F_xFeAs$". *Nature* **453** (7193): 376–378. Bibcode:2008Natur.453..376T. doi:10.1038/nature06972. PMID 18432191.

[45] Adrian Cho. "Second Family of High-Temperature Superconductors Discovered". ScienceNOW Daily News.

[46] Zhi-An Ren; et al. (2008). "Superconductivity and phase diagram in iron-based arsenic-oxides ReFeAsO1-d (Re = rare-earth metal) without fluorine doping". *EPL* **83**: 17002. arXiv:0804.2582. Bibcode:2008EL.....8317002R. doi:10.1209/0295-5075/83/17002.

[47] Li, Yinwei; Hao, Jian; Liu, Hanyu; Li, Yanling; Ma, Yanming (2014-05-07). "The metallization and superconductivity of dense hydrogen sulfide". *The Journal of Chemical Physics* **140** (17): 174712. doi:10.1063/1.4874158. ISSN 0021-9606.

[48] Drozdov, A. P.; Eremets, M. I.; Troyan, I. A.; Ksenofontov, V.; Shylin, S. I. (2015). "Conventional superconductivity at 203 kelvin at high pressures in the sulfur hydride system". *Nature* **525** (7567): 73–6. doi:10.1038/nature14964. ISSN 0028-0836. PMID 26280333.

[49] Islam et al, *A review of offshore wind turbine nacelle: Technical challenges, and research and developmental trends*. In: *Renewable and Sustainable Energy Reviews* 33, (2014), 161–176, doi:10.1016/j.rser.2014.01.085

[50] Linder, Jacob; Robinson, Jason W. A. (2 April 2015). "Superconducting spintronics". *Nature Physics* **11** (4): 307–315. doi:10.1038/nphys3242.

[51] Thomas et al, *Superconducting transmission lines – Sustainable electric energy transfer with higher public acceptance?* In: *Renewable and Sustainable Energy Reviews* 55, (2016), 59–72, doi:10.1016/j.rser.2015.10.041.

[52] "All Nobel Prizes in Physics". *Nobelprize.org*. Nobel Media AB 2014.

18.9   Further reading

- Hagen Kleinert (1989). "Superflow and Vortex Lines". *Gauge Fields in Condensed Matter* **1**. World Scientific. ISBN 9971-5-0210-0.

- Anatoly Larkin; Andrei Varlamov (2005). *Theory of Fluctuations in Superconductors*. Oxford University Press. ISBN 0-19-852815-9.

- A. G. Lebed (2008). *The Physics of Organic Superconductors and Conductors* **110** (1st ed.). Springer. ISBN 978-3-540-76667-4.

- Jean Matricon; Georges Waysand; Charles Glashausser (2003). *The Cold Wars: A History of Superconductivity*. Rutgers University Press. ISBN 0-8135-3295-7.

- "Physicist Discovers Exotic Superconductivity". ScienceDaily. 17 August 2006.

- Michael Tinkham (2004). *Introduction to Superconductivity* (2nd ed.). Dover Books. ISBN 0-486-43503-2.

- Terry Orlando; Kevin Delin (1991). *Foundations of Applied Superconductivity*. Prentice Hall. ISBN 978-0-201-18323-8.

- Paul Tipler; Ralph Llewellyn (2002). *Modern Physics* (4th ed.). W. H. Freeman. ISBN 0-7167-4345-0.

18.10   External links

- Everything about superconductivity: properties, research, applications with videos, animations, games

- Video about Type I Superconductors: R=0/transition temperatures/ B is a state variable/ Meissner effect/ Energy gap(Giaever)/ BCS model

- Superconductivity: Current in a Cape and Thermal Tights. An introduction to the topic for non-scientists National High Magnetic Field Laboratory

- Lectures on Superconductivity (series of videos, including interviews with leading experts)

- Superconductivity News Update

- Superconductor Week Newsletter – industry news, links, et cetera

- Superconducting Magnetic Levitation

- National Superconducting Cyclotron Laboratory at Michigan State University

- YouTube Video Levitating magnet

- International Workshop on superconductivity in Diamond and Related Materials (free download papers)

- New Diamond and Frontier Carbon Technology Volume 17, No.1 Special Issue on Superconductivity in CVD Diamond

- DoITPoMS Teaching and Learning Package – "Superconductivity"

- The Nobel Prize for Physics, 1901–2008

- folding hands-on activities about superconductivity

Chapter 19

Superheated water

Superheated water is liquid water under pressure at temperatures between the usual boiling point, 100 °C (212 °F) and the critical temperature, 374 °C (705 °F). It is also known as "subcritical water" or "pressurized hot water." Superheated water is stable because of overpressure that raises the boiling point, or by heating it in a sealed vessel with a headspace, where the liquid water is in equilibrium with vapour at the saturated vapor pressure. This is distinct from the use of the term superheating to refer to water at atmospheric pressure above its normal boiling point, which has not boiled due to a lack of nucleation sites (sometimes experienced by heating liquids in a microwave).

Many of water's anomalous properties are due to very strong hydrogen bonding. Over the superheated temperature range the hydrogen bonds break, changing the properties more than usually expected by increasing temperature alone. Water becomes less polar and behaves more like an organic solvent such as methanol or ethanol. Solubility of organic materials and gases increases by several orders of magnitude and the water itself can act as a solvent, reagent and catalyst in industrial and analytical applications, including extraction, chemical reactions and cleaning.

19.1 Change of properties with temperature

All materials change with temperature, but water exhibits greater changes than would be expected from temperature considerations alone. Viscosity and surface tension of water drop and diffusivity increases with increasing temperature. [1] Self-ionization of water increases with temperature, and the pKw of water at 250 °C is closer to 11 than the more familiar 14 at 25 °C. This means the concentration of hydronium ion (H
$3O+$
) is higher, although the level of hydroxide (OH–
) is increased by the same amount so the pH is still neutral. Specific heat capacity at constant pressure also increases with temperature, from 4.187 kJ/kg at 25 °C to 8.138 kJ/kg at 350 °C. A significant effect on the behaviour of water at high temperatures is decreased dielectric constant (relative permittivity).[2]

19.2 Explanation of anomalous behaviour

Water is a polar molecule, where the centers of positive and negative charge are separated; so molecules will align with an electric field. The extensive hydrogen bonded network in water tends to oppose this alignment, and the degree of alignment is measured by the relative permittivity. Water has a high relative permittivity of about 80 at room temperature; because polarity shifts are rapidly transmitted through shifts in orientation of the linked hydrogen bonds. This allows water to dissolve salts, as the attractive electric field between ions is reduced by about 80–fold.[1] Thermal motion of the molecules disrupts the hydrogen bonding network as temperature increases; so relative permittivity decreases with temperature to about 7 at the critical temperature. At 205 °C the relative permittivity falls to 33, the same as methanol at room temperature. Thus water behaves like a water / methanol mixture between 100 °C and 200 °C. Disruption of extended

Pressure cookers produce superheated water, which cooks the food much more rapidly than boiling water.

hydrogen bonding allows molecules to move more freely (viscosity, diffusion and surface tension effects), and extra energy must be supplied to break the bonds (increased heat capacity).

19.3   Solubility

19.3.1   Organic compounds

Organic molecules often show a dramatic increase in solubility with temperature, partly because of the polarity changes described above, and also because the solubility of sparingly soluble materials tends to increase with temperature as they have a high enthalpy of solution. Thus materials generally considered "insoluble" can become soluble in superheated water. E.g., the solubility of PAHs is increased by 5 orders of magnitude from 25 °C to 225 °C[3] and naphthalene, for example, forms a 10% wt solution in water at 270 °C, and the solubility of the pesticide chloranthonil with temperature is shown in the table below.[2]

Thus superheated water can be used to process many organic compounds with significant environmental benefits compared to the use of conventional organic solvents.

19.3.2 Salts

Despite the reduction in relative permittivity, many salts remain soluble in superheated water until the critical point is approached. Sodium chloride, for example, dissolves at 37 wt% at 300 °C[4] As the critical point is approached, solubility drops markedly to a few ppm, and salts are hardly soluble in supercritical water. Some salts show a reduction in solubility with temperature, but this behaviour is less common.

19.3.3 Gases

The solubility of gases in water is usually thought to decrease with temperature, but this only occurs to a certain temperature, before increasing again. For nitrogen, this minimum is 74 °C and for oxygen it is 94 °C[5] Gases are soluble in superheated water at elevated pressures. Above the critical temperature, water is completely miscible with all gasses. The increasing solubility of oxygen in particular allows superheated water to be used for wet oxidation processes.

19.4   Corrosion

Superheated water can be more corrosive than water at ordinary temperatures, and at temperatures above 300 °C special corrosion resistant alloys may be required, depending on other dissolved components. Continuous use of carbon steel pipes for 20 years at 282 °C has been reported without significant corrosion,[6] and stainless steel cells showed only slight deterioration after 40-50 uses at temperatures up to 350 °C.[7] The degree of corrosion that can be tolerated depends on the use, and even corrosion resistant alloys can fail eventually. Corrosion of an Inconel U-tube in a heat exchanger was blamed for an accident at a nuclear power station.[8] Therefore, for occasional or experimental use, ordinary grades of stainless steel are probably adequate with continuous monitoring, but for critical applications and difficult to service parts, extra care needs to be taken in materials selection.

19.5   Effect of pressure

At temperatures below 300 °C water is fairly incompressible, which means that pressure has little effect on the physical properties of water, provided it is sufficient to maintain a liquid state. This pressure is given by the saturated vapour pressure, and can be looked up in steam tables, or calculated.[9] As a guide, the saturated vapour pressure at 121 °C is 200 kPa, 150 °C is 470 kPa, and 200 °C is 1,550 kPa. The critical point is 21.7 MPa at a temperature of 374 °C, above which water is supercritical rather than superheated. Above about 300 °C, water starts to behave as a near-critical liquid, and physical properties such as density start to change more significantly with pressure. However, higher pressures increase the rate of extractions using superheated water below 300 °C. This could be due to effects on the substrate, particularly plant materials, rather than changing water properties.

19.6   Energy requirements

The energy required to heat water is significantly lower than that needed to vaporize it, for example for steam distillation[10] and the energy is easier to recycle using heat exchangers. The energy requirements can be calculated from steam tables. For example, to heat water from 25 °C to steam at 250 °C at 1 atm requires 2869 kJ/kg. To heat water at 25 °C to liquid water at 250 °C at 5 MPa requires only 976 kJ/kg. It is also possible to recover much of the heat (say 75%)

from superheated water, and therefore energy use for superheated water extraction is less than one sixth that needed for steam distillation. This also means that the energy contained in superheated water is insufficient to vaporise the water on decompression. In the above example, only 30% of the water would be converted to vapour on decompression from 5 MPa to atmospheric pressure.[2]

19.7   Extraction

Extraction using superheated water tends to be fast because diffusion rates increase with temperature. Organic materials tend to increase in solubility with temperature, but not all at the same rate. For example, in extraction of essential oils from rosemary[11] and coriander,[12] the more valuable oxygenated terpenes were extracted much faster than the hydrocarbons. Therefore, extraction with superheated water can be both selective and rapid, and has been used to fractionate diesel and woodsmoke particulates.[13] Superheated water is being used commercially to extract starch material from marsh mallow root for skincare applications[14] and to remove low levels of metals from a high-temperature resistant polymer.[15][16]

For analytical purposes, superheated water can replace organic solvents in many applications, for example extraction of PAH's from soils[17] and can also be used on a large scale to remediate contaminated soils, by either extraction alone or extraction linked to supercritical or wet oxidation.[18]

19.8   Reactions

Superheated water, along with supercritical water, has been used to oxidise hazardous material in the wet oxidation process. Organic compounds are rapidly oxidised without the production of toxic materials sometimes produced by combustion. However, when the oxygen levels are lower, organic compounds can be quite stable in superheated water. As the concentration of hydronium (H
$3O+$
) and hydroxide (OH−
) ions are 100 times larger than in water at 25 °C, superheated water can act as a stronger acid and a stronger base, and many different types of reaction can be carried out. An example of a selective reaction is oxidation of ethylbenzene to acetophenone, with no evidence of formation of phenylethanoic acid, or of pyrolysis products.[7] Several different types of reaction in which water was behaving as reactant, catalyst and solvent were described by Katritzky et al.[19] Triglycerides can be hydrolysed to free fatty acids and glycerol by superheated water at 275 °C,[20] which can be the first in a two-stage process to make biodiesel. [21] Superheated water can be used to chemically convert organic material into fuel products. This is known by several terms, including direct hydrothermal liquefaction,[22] and hydrous pyrolysis. A few commercial scale applications exist. Thermal depolymerization or thermal conversion (TCC) uses superheated water at about 250 °C to convert turkey waste into a light fuel oil and is said to process 200 tons of low grade waste into fuel oil a day. [23] The initial product from the hydrolysis reaction is de-watered and further processed by dry cracking at 500 °C. The "SlurryCarb" process operated by EnerTech uses similar technology to decarboxylate wet solid biowaste, which can then be physically dewatered and used as a solid fuel called E-Fuel. The plant at Rialto is said to be able to process 683 tons of waste per day. [24] The HTU or Hydro Thermal Upgrading process appears similar to the first stage of the TCC process. A demonstration plant is due to start up in The Netherlands said to be capable of processing 64 tons of biomass (dry basis) per day into oil.[25]

19.9   Chromatography

Reverse phased HPLC often uses methanol / water mixtures as the mobile phase. Since the polarity of water spans the same range from 25 to 205 °C, a temperature gradient can be used to effect similar separations, for example of phenols. [26] The use of water allows the use of the flame ionisation detector (FID), which gives mass sensitive output for nearly all organic compounds. [27] The maximum temperature is limited to that at which the stationary phase is stable. C18 bonded phases which are common in HPLC seem to be stable at temperatures up to 200 °C, far above that of pure silica, and polymeric styrene / divinylbenzene phases offer similar temperature stability. [28] Water is also compatible with use of an ultraviolet detector down to a wavelength of 190 nm.

19.10   See also

- Superheated steam
- Pressurized water reactor
- Super critical carbon dioxide

19.11   References

[1] Chaplin, Martin (2008-01-04). "Explanation of the physical anomalies of water". London South Bank University. Retrieved 2008-01-15.

[2] Clifford, A.A. (2008-01-04). "Changes of water properties with temperature". Retrieved 2008-01-15.

[3] Miller, D.J.; Hawthorne, S.B; Gizir, A.M.; Clifford, A.A. (1998). "Solubility of polycyclic aromatic hydrocarbons in subcritical water from 298 K to 498 K". *Journal of Chemical Engineering Data* (American Chemical Society) 43 (6): 1043–1047. doi:10.1021/je980094g. Retrieved 2008-01-14.

[4] Letcher, Trevor M. (2007). *Thermodynamics, solubility and environmental issues*. Elsevier. p. 60. ISBN 0-444-52707-9.

[5] "Guideline on the Henry's constant and vapor-liquid distribution constant for gases in H2O and D2O at high temperatures" (PDF). International Association for the Properties of Water and Steam. September 2004. Retrieved 2008-01-14.

[6] Burnham, Robert N.; et al. (2001). "Measurement of the flow of superheated water in blowdown pipes at MP2 using an ultrasonic clamp-on method" (PDF). Panametrix. Retrieved 2008-01-14.

[7] Holliday, Russel L.; Yong, B.Y.M.; Kolis, J.W. (1998). "Organic synthesis in subcritical water. Oxidation of alkyl aromatics". *Journal of Supercritical Fluids* (Elsevier) 12 (3): 255–260. doi:10.1016/S0896-8446(98)00084-9. Retrieved 2008-01-12.

[8] "Corrosion seen as A-plant accident cause". New York Times. 2000-03-03. Retrieved 2008-01-15.

[9] Clifford, A.A. (2007-12-04). "Superheated water: more details". Retrieved 2008-01-12.

[10] King, Jerry W. "Poster 12. Pressurized water extraction: resources and techniques for optimizing analytical applications, Image 13". Los Alamos National Laboratories. Retrieved 2008-01-12.

[11] Basile, A.; et al. (1998). "Extraction of Rosemary by Superheated Water". *J. Agric. Food Chem.* (American Chemical Society) 46 (12): 5205–5209. doi:10.1021/jf980437e. Retrieved 2008-01-12.

[12] Eikani, M.H.; Golmohammad, F.; Rowshanzamir, S. "Subcritical water extraction of essential oils from coriander seeds (Corianrum sativum L.)" (PDF). Retrieved 2008-01-04.

[13] Kubatova, Alena; Mayia Fernandez; Steven Hawthorne (2002-04-09). "A new approach to characterizing organic aerosol (wood smoke and diesel exhaust particulate) using subcritical water fractionation" (PDF). *PM2.5 and electric power generation: recent findings and implications*. Pittsburgh, PA: National Energy Technology Laboratory. Retrieved 2008-01-10.

[14] "LINK Competitive Industrial Materials from Non-Food Crops Applications: water and superheated water" (PDF). *Newsletter No.8*. BBSRC. Spring 2007. Retrieved 2008-01-08.

[15] Clifford, A.A. (2007-12-04). "Applications: water and superheated water". Retrieved 2008-01-08.

[16] Clifford, Tony (Nov 5–8, 2006). "Separations using superheated water". *8th International Symposium on Supercritical Fluids*. Kyoto, Japan. Retrieved 2008-01-16.

[17] Kipp, Sabine; et al. (July 1998). "Coupling superheated water extraction with enzyme immunoassay for an efficient and fast PAH screening in soil". *Talanta* (Elsevier Science BV) 46 (3): 385–393. doi:10.1016/S0039-9140(97)00404-9. PMID 18967160. Retrieved 2008-01-12.

[18] Hartonen, K; Kronholm and Reikkola (2005). Jalkanen, Anneli; Nygren, Pekka, eds. *Sustainable use of renewable natural resources – principles and practice* (PDF). Chapter 5.2 Utilisation of high temperature water in the purification of water and soil: University of Helsinki Department of Forest Ecology. ISBN 952-10-2817-3.

[19] Katritzki, A.R.; S. M. Allin; M. Siskin (1996). "Aquathermolysis: reaction of organic compounds with superheated water" (PDF). *Accounts of Chemical Research* (American Chemical Society) **29** (8): 399–406. doi:10.1021/ar950144w. Retrieved 2008-01-14.

[20] King, Jerry W.; Holliday, R.L.; List, G.R. (December 1999). "Hydrolysis of soubean oil in a subcritical water flow reactor". *Green Chemistry* (Royal Society of Chemistry): 261–264. Retrieved 2008-01-12.

[21] Saka, Shiro; Kusdiana, Dadan. "NEDO "High efficiency bioenergy conversion project"R & D for biodiesel fuel (BDF) by two step supercritical methanol method" (PDF). Retrieved 2008-01-12.

[22] "Biomass Program, direct Hydrothermal Liquefaction". US Department of Energy. Energy Efficiency and Renewable Energy. 2005-10-13. Archived from the original on 2008-01-03. Retrieved 2008-01-12.

[23] "About TCP Technology". Renewable Environmental Solutions LLC. Retrieved 2008-01-12.

[24] Sforza, Teri (2007-03-14). "New plan replaces sewage sludge fiasco". Orange county register. Retrieved 2008-01-27.

[25] Goudriaan, Frans; Naber, Jaap and van den Berg, Ed. "Conversion of Biomass Residues to Transportation Fuels with th HTU Process" (PDF). Retrieved 2008-01-12. Cite uses deprecated parameter |coauthors= (help)

[26] Yarita, Takashi; Nakajima, R.; Shibukawa, M. (February 2003). "Superheated water chromatography of phenols using poly(styrene-divinylbenzene) packings as a stationary phase". *Analytical Sciences* (The Japan Society for Analytical Chemistry) **19** (2): 269–272. doi:10.2116/analsci.19.269. PMID 12608758. Retrieved 2008-01-12.

[27] Smith, Roger; Young, E.; Sharp, B. "Superheated water chromatography – flame ionization detection" (PDF). Retrieved 2001-01-12. Check date values in: |access-date= (help)

[28] Smith, R. M.; Burgess, R.J. (1996). "Superheated water – a clean eluent for reverse phase high performance chromatography". *Analytical Communications* (Royal Society of Chemistry) **33** (9): 327–329. doi:10.1039/AC9963300327. Retrieved 2008-01-12.

19.12 External links

- The International Association for the Properties of Water and Steam
- Calculator for vapour pressure and enthalpy of superheated water.

Chapter 20

Superheating

This article is about the phenomenon where a liquid can exist in a metastable state above its boiling point. See superheated water for pressurized water above 100 °C. See superheater for the device used in steam engines.

In physics, **superheating** (sometimes referred to as **boiling retardation**, or **boiling delay**) is the phenomenon in which a liquid is heated to a temperature higher than its boiling point, without boiling. Superheating is achieved by heating a homogeneous substance in a clean container, free of nucleation sites, while taking care not to disturb the liquid.

20.1 Cause

Water is said to "boil" when bubbles of water vapor grow without bound, bursting at the surface. For a vapor bubble to expand, the temperature must be high enough that the vapor pressure exceeds the ambient pressure (the atmospheric pressure, primarily). Below that temperature, a water vapor bubble will shrink and vanish.

Superheating is an exception to this simple rule; a liquid is sometimes observed not to boil even though its vapor pressure does exceed the ambient pressure. The cause is an additional force, the surface tension, which suppresses the growth of bubbles.[1]

Surface tension makes the bubble act like a rubber balloon (more precisely, one that is under-inflated so that the rubber is still elastic). The pressure inside is raised slightly by the "skin" attempting to contract. For the bubble to expand, the temperature must be raised slightly above the boiling point to generate enough vapor pressure to overcome both surface tension and ambient pressure.

What makes superheating so explosive is that a larger bubble is easier to inflate than a small one; just as when blowing up a balloon, the hardest part is getting started. It turns out the excess pressure due to surface tension is inversely proportional to the diameter of the bubble.[2] This means if the largest bubbles in a container are only a few micrometres in diameter, overcoming the surface tension may require exceeding the boiling point by several degrees Celsius. Once a bubble does begin to grow, the pressure due to the surface tension reduces, so it expands explosively. In practice, most containers have scratches or other imperfections which trap pockets of air that provide starting bubbles. But a container of liquid with only microscopic bubbles can superheat dramatically.

20.2 Occurrence via microwave oven

Superheating can occur when an undisturbed container of water is heated in a microwave oven. When the container is removed, the water still appears to be below the boiling point. However, once the water is disturbed, some of it violently flashes to steam, potentially spraying boiling water out of the container.[3] The boiling can be triggered by jostling the cup, inserting a stirring device, or adding a substance like instant coffee or sugar. The chances of superheating are greater with smooth containers, because scratches or chips can house small pockets of air, which serve as nucleation points. Chances

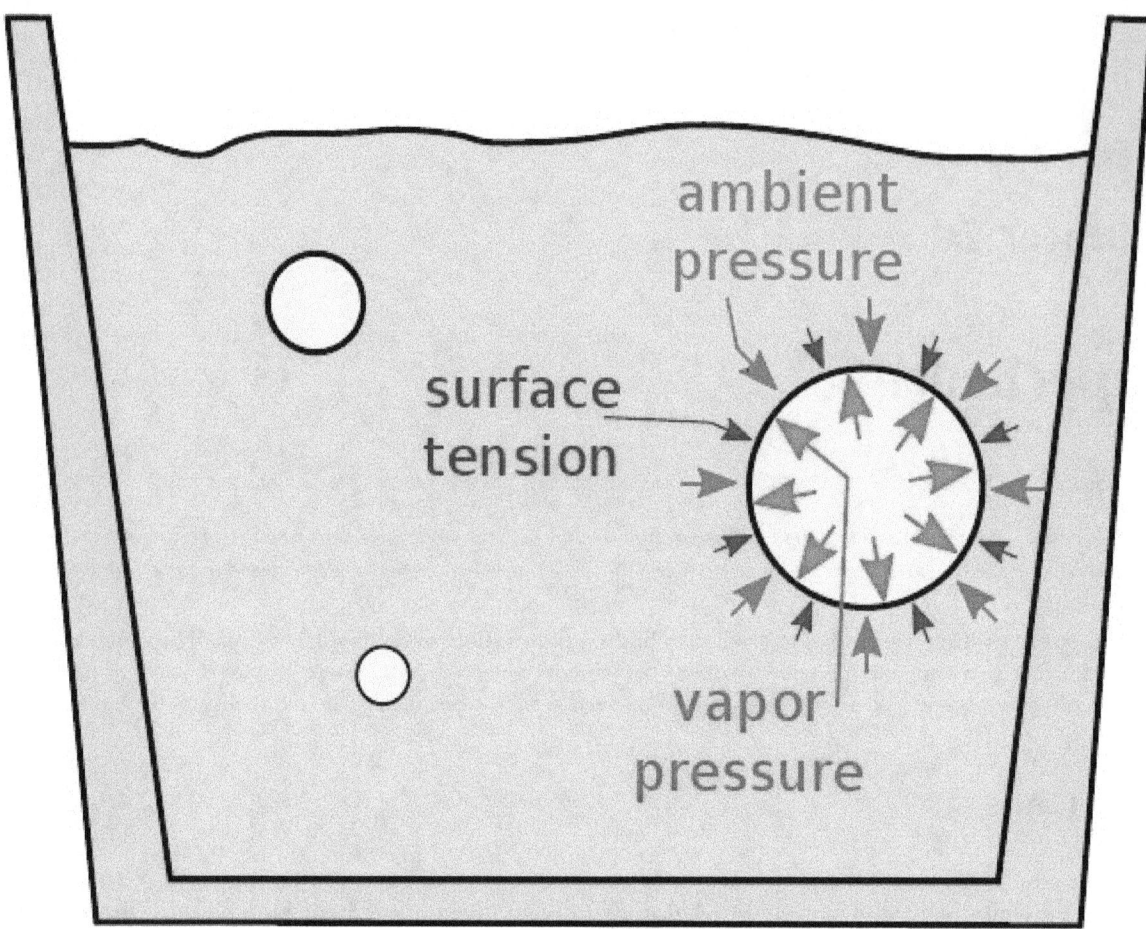

For boiling to occur, the vapor pressure must exceed the ambient pressure plus a small amount of pressure induced by surface tension

of superheating can increase with repeated heating and cooling cycles of an undisturbed container, like when a forgotten coffee cup is re-heated without being removed from a microwave oven. This is due to heating cycles progressively degassing the liquid. There are ways to prevent superheating in a microwave oven, such as putting a popsicle stick in the glass or using a scratched container.

20.3   Applications

Superheated hydrogen liquid is used in bubble chambers.

20.4   Myth

There is a common belief that superheating can occur only in pure substances. This is untrue, as superheating has been observed in coffee and other impure liquids. Impurities do prevent superheating if they introduce nucleation sites (rough areas where gas is trapped); for example, sand tends to suppress superheating in water. Dissolved gas can also provide nucleation sites when it comes out of solution and forms bubbles. However, an impurity such as salt or sugar, dissolved in water to form a homogeneous solution, does not prevent superheating. Other liquids are known to superheat including 2% milk and almond milk.[4]

20.5 See also

- Autoclave

- Bumping (chemistry)

- Critical point (thermodynamics)

- Supercooling

- Supersaturation

- Subcooling

20.6 References

[1] Critical Droplets and Nucleation, Cornell Solid State Lab

[2] Atmosphere-ocean Interaction By Eric Bradshaw Kraus, Joost A. Businger Published by Oxford University Press US, 1994 ISBN 0-19-506618-9, pg 60.

[3] Urban Legends Reference Pages: Superheated Microwaved Water

[4] Beaty, William & U. Washington. "Impure water can also undergo superheating. Any superheated liquid can be dangerous so when superheating exercise caution". Retrieved 2007-11-24.

20.7 External links

- Video of superheated water in a microwave explosively flash boiling, why it happens, and why it's dangerous.

- Bloomfield, Louis A. "A series of superheated water with oil film experiments done in the microwave by Louis A. Bloomfield, physics professor at the University of Virginia. Experiment #13 proceeds with surprising violence". Archived from the original on 2 June 2008.

- Video of superheated water in a pot.

Chapter 21

Thermo-dielectric effect

The **thermo-dielectric effect** is the production of electric currents and charge separation during phase transition.

This interesting effect was discovered by Joaquim da Costa Ribeiro in 1944. The Brazilian physicist observed that solidification and melting of many dielectrics are accompanied by charge separation. A thermo-dielectric effect was demonstrated with carnauba wax, naphthalene and paraffin. Charge separation in ice was also expected. This effect was observed during water freezing period, electrical storm effects can be caused by this strange phenomenon. Effect was measured by many researches - Bernhard Gross, Armando Dias Tavares, Sergio Mascarenhas etc. César Lattes (co-discoverer of the pion) supposed that this was the only effect ever to be discovered entirely in Brasil.

21.1 References

- Gross, B. (1954). "Theory of Thermodielectric eff ect". *Physical Review* **94** (6): 1545–1551. Bibcode:1954 doi:10.1103/PhysRev.94.1545.

- Saldaña, Juan José (2006). *Science in Latin America: a history*. ISBN 978-0-292-71271-3.

21.2 External links

- Thermodielectric effect measurement

21.3 Text and image sources, contributors, and licenses

21.3.1 Text

- **State of matter** *Source:* https://en.wikipedia.org/wiki/State_of_matter?oldid=698119776 *Contributors:* CYD, Zardoz, Olof, Zundark, William Avery, Heron, Ubiquity, Ixfd64, Docu, Julesd, Seani, Reddi, Alexfiles, Robbot, Mayooranathan, Academic Challenger, Pepijn Schmitz, Rursus, Giftlite, Smjg, Art Carlson, Jackol, SoWhy, Rdsmith4, Oneiros, JimWae, DragonflySixtyseven, Glogger, ArthurDenture, Mike Rosoft, Discospinster, Rich Farmbrough, Vsmith, Walden, Rgdboer, Shanes, Smalljim, Zetawoof, Nsaa, Alansohn, Hydriotaphia, Wtmitchell, Mikeo, Vuo, Kazvorpal, Oleg Alexandrov, LOL, Pol098, Ruud Koot, MONGO, Yuriybrisk, Rjwilmsi, DoubleBlue, ElfQrin, Srleffler, Kri, The-Sun, Imnotminkus, King of Hearts, Moocha, DVdm, Mercury McKinnon, Wavelength, Phantomsteve, Jeffhoy, Arado, Madkayaker, Gaius Cornelius, CambridgeBayWeather, Pseudomonas, NawlinWiki, Dhollm, BOT-Superzerocool, Wknight94, 2over0, Closedmouth, Pb30, Willtron, Yonir, Moomoomoo, JDspeeder1, Cmglee, Knowledgeum, SmackBot, McGeddon, Edgar181, Skizzik, Quinsareth, Miquonranger03, Colonies Chris, Darth Panda, Addshore, RedHillian, Ghiraddje, Valenciano, Bombshell, Keramos, Kipala, A. Parrot, Beetstra, Cratylus3, Mark999, BranStark, Iridescent, Joseph Solis in Australia, Newone, Igoldste, RekishiEJ, Courcelles, Tltltetd, Dycedarg, MarsRover, Cydebot, Mato, Xxanthippe, Bazzargh, Christian75, DumbBOT, Mtpaley, Epbr123, Urdna, N5iln, Headbomb, Kathovo, James086, Yettie0711, PJtP, Escarbot, Mentifisto, Daniels220, Jayron32, Jj137, Gökhan, JAnDbot, D99figge, YK Times, Bongwarrior, VoABot II, JNW, Father Goose, Rich257, Animum, Dirac66, Philg88, Khalid Mahmood, FisherQueen, Anaxial, Sm8900, Jonathan Hall, R'n'B, CommonsDelinker, Wiki Raja, Mausy5043, J.delanoy, Pharaoh of the Wizards, MITBeaverRocks, Uncle Dick, BobEnyart, Victuallers, Myrin1, Gombang, Chemicalrubber, NewEnglandYankee, Kraftlos, Bonadea, Ja 62, CrazyRob926, Useight, Martial75, Funandtrvl, CWii, ABF, Flyingidiot, Jeff G., Indubitably, Barneca, Philip Trueman, TXiKiBoT, Monkey Bounce, Dendodge, Corvus cornix, Cremepuff222, RiverStyx23, Venny85, Madhero88, Blurpeace, Falcon8765, Pageman~enwiki, Brianga, Universaladdress, HPeugeot~enwiki, NHRHS2010, Cryonic07, SieBot, Alessgrimal, Nubiatech, Euryalus, Paradoctor, Winchelsea, Yintan, Happysailor, Flyer22 Reborn, Socal gal at heart, TrufflesTheLamb, Dillard421, Svick, Mygerardromance, Pinkadelica, Denisarona, Escape Orbit, Kanonkas, ShajiA, Troy 07, Ainlina, Faithlessthewonderboy, Elassint, ClueBot, Wookie501, Rumping, GorillaWarfare, Snigbrook, The Thing That Should Not Be, Ariadacapo, Tizeff, Polyamorph, GoEThe, Dylan620, Mad.martian999, Prognitor~enwiki, Quantumspinhall, Skihatboatbike, Djr32, Excirial, Alexbot, Coralmizu, Ykhwong, Pot, Singhalawap, Dekisugi, The Red, Oswald07, JasonAQuest, Mikhailov Kusserow, Thingg, Horselover Frost, PCHS-NJROTC, Burner0718, SoxBot III, Jmanigold, XLinkBot, Spitfire, PseudoOne, Leftspk, Andypandy2020, Saeed.Veradi, Libcub, Skarebo, SilvonenBot, Noctibus, JinJian, ZooFari, Thatguyflint, HexaChord, IBeAiMpErS0naT0r, Xvijayx, Addbot, Xp54321, Some jerk on the Internet, Jojhutton, PaterMcFly, Ronhjones, TutterMouse, Fieldday-sunday, Bertrc, Icelazer3, Shirtwaist, Vishnava, CanadianLinuxUser, NjardarBot, Download, PranksterTurtle, Glane23, Chzz, FCSundae, Kyle1278, CuteHappyBrute, Eh kia, Tide rolls, Gail, Micki, Cesaar, Angrysockhop, Legobot, Luckas-bot, Yobot, Dr. Footfoe, 2D, Andreasmperu, II MusLiM HyBRiD II, THEN WHO WAS PHONE?, Empireheart, IW.HG, Tempodivalse, Nintend06, Quangbao, Hairhorn, Daniele Pugliesi, Jim1138, Lewismith3, Piano non troppo, Kingpin13, Dinesh smita, Westonpark, Materialscientist, Spirit469, ImperatorExercitus, 90 Auto, Citation bot, OllieFury, ArthurBot, Xqbot, TinucherianBot II, Beaserbebbeb, Pvkeller, Grim23, GrouchoBot, Charliekarst, Hamhat, Backpackadam, Amaury, Logger9, JediMaster362, Wasteman1066, Lagooncopperorange, FrescoBot, LucienBOT, Lothar von Richthofen, Hielor, Cannolis, Citation bot 1, Athenanoenvoy, Pinethicket, Boulaur, Overthinkingly, Tom.Reding, ContinueWithCaution, Johann137, Jauhienij, Gamewizard71, FoxBot, Yunshui, Zvn, January, Allen4names, Theo10011, Reaper Eternal, Jeffrd10, Tbhotch, Hornlitz, Marie Poise, DARTH SIDIOUS 2, NerdyScienceDude, Chuanenlin123, Jack Schlederer, Kenvancleve, WikitanvirBot, Ajraddatz, DillonLMcCabe, Mordgier, Katherine, Gcastellanos, The Sharminator, Tommy2010, Wikipelli, John Cline, Zane45177, Alpha Quadrant (alt), DidgeGuy, AndrewN, Yamumyadad, Jay-Sebastos, Jesanj, Brandmeister, Donner60, Ashunigam, Orange Suede Sofa, ChuispastonBot, Llightex, ShatteredSpiral, Isocliff, ClueBot NG, Gareth Griffith-Jones, MelbourneStar, UAK32, Rtucker913, Yourmomblah, Theimmaculatechemist, 123Hedgehog456, Whazie, Lfmm11, Mtheodric, Kumamah, جاهر ما جاهر ماس, Friend20gan, MerllwBot, Helpful Pixie Bot, Candleabracadabra, Bibcode Bot, Mingmingla, Mark Fole, BZTMPS, Lowercase sigmabot, Yetisyny, Hallows AG, MusikAnimal, Frze, AvocatoBot, Planetary Chaos Redux, Tatchell, Nsda, ChE Fundamentalist, Ugncreative Usergname, Joydeep, Averysamantha, Snow Blizzard, ImhotepBallZ, Mantike, Kydon Shadow, Th4n3r, The Illusive Man, ChrisGualtieri, EuroCarGT, Sidsandyy, Lugia2453, CaSJer, Frosty, SFK2, Jamesx12345, Perfecttwoegan, Reatlas, Passengerpigeon, Epicgenius, I am One of Many, Lsmll, Surfer43, Tentinator, Cebr1979, Amdz96, Backendgaming, Lince celeste, DavidLeighEllis, Ginsuloft, WikiMannWikiMann, Annieminnie21, Rickdutta, Meteor sandwich yum, Muneeb abdelhadi, ShahryarAhmad27900, Nijuthomasgeorge, Bilorv, Saaaaaaaaaa, Ookami-to-Koneko, J73364, S r u fejvd u ub creating u of, Amortias, Natdeso, Adenine2k, Lalalalalala is too great, ElmoLover88545, Darkmatter1435, Chetan082, Alango1998, XxX$pace5Xxx, Cambles13, Johno2000, Jsbdjejnwn, Weegeerunner, Slayeredwarrior, Account50, Blissy2005, Minnie431, KasparBot and Anonymous: 823

- **Enthalpy of fusion** *Source:* https://en.wikipedia.org/wiki/Enthalpy_of_fusion?oldid=695619922 *Contributors:* Mav, Bryan Derksen, Heron, Fred Bauder, Mahjongg, Kku, Greatpatton, Bjh21, Wampa Jabba, Fvw, Donarreiskoffer, Gentgeen, Michael Devore, Eequor, Dreamtheater, Discospinster, Rich Farmbrough, Vsmith, ArnoldReinhold, Femto, RAM, Mac Davis, Walkerma, Wdfarmer, Stephan Leeds, Gene Nygaard, LimoWreck, Graham87, V8rik, Joffan, Physchim62, Chobot, Gwernol, Alethiareg, Dhollm, E2mb0t~enwiki, Jobаннő, Melchoir, Nmulder, Isaac Dupree, Bluebot, MalafayaBot, DHN-bot~enwiki, Colonies Chris, Bekom, Greg5030, Drc79, Mbeychok, Pierre cb, Ryulong, Angelpeream, Nutster, Stifynsemons, Mikiemike, Kaldosh, Christian75, Epbr123, Escarbot, JAnDbot, DuncanHill, Pi.1415926535, User A1, Joseph.Kidder, MartinBot, Krushdiva, Thermbal, Hodja Nasreddin, Nothingofwater, Andareed, Fiver2552, WilfriedC, Towerman86, Apg 77, CardinalDan, Idioma-bot, VolkovBot, Rogator, Rei-bot, Cosmium, Bitjungle, Ryanmoffet, NonLocalYokel, Cfgammie, Kunal chawla09, Infinitegoof, Neparis, Kbrose, SieBot, ClueBot, Seervoitek, Karrade247, ChandlerMapBot, Djr32, Banano03, Lambtron, Fastily, Dezaxa, Avoided, SilvonenBot, Monfornot, Addbot, Xorxar, WikiUserPedia, Xicer9, Ehrenkater, Luckas-bot, Enisbayramoglu, Materialscientist, ArthurBot, زادگان یلق, Amaury, Erik9, LucienBOT, Jesterbard, Gamewizard71, MusicScienceGuy, Slaviboy, DD5baller, Josve05a, Cobaltcigs, ChemMater, ClueBot NG, Frietjes, Marostyle, ThatAMan, Gredner, JiriMatejicek, Mgibby5, Abi2013, TheLuigiBrother, WardtMed-DevRD, Yazhini Tamilanban, Rmmilewi, Ahiijny, Vpab15, Clauds526 and Anonymous: 152

- **Enthalpy of sublimation** *Source:* https://en.wikipedia.org/wiki/Enthalpy_of_sublimation?oldid=681160370 *Contributors:* Jdpipe, Physchim62, Bhny, Dhollm, SmackBot, Ryulong, Ioannes Pragensis, Gamma Guy, Runch, Headbomb, Alexbot, PixelBot, Addbot, Daniele Pugliesi, یلق زادگان, Erik9bot, Double sharp, ZéroBot, Mn-imhotep, CarrieVS, Matin.ahmadpor and Anonymous: 2

- **Enthalpy of vaporization** *Source:* https://en.wikipedia.org/wiki/Enthalpy_of_vaporization?oldid=694423115 *Contributors:* Magnus Manske, Mav, Jeronimo, Andre Engels, William Avery, SimonP, Jdpipe, Ewen, Bdesham, Patrick, Pit~enwiki, Fred Bauder, Kku, Guerby, Ahoerstemeier, Docu, Julesd, Glenn, HoIIgor, Malbi, Greatpatton, Zoicon5, Francs2000, Donarreiskoffer, Gentgeen, Robbot, Hankwang, HaeB, Buster2058, Andries, Herbee, Brockert, Bobblewik, Icairns, NathanHurst, Vsmith, Hibisco-da-Ilha~enwiki, Femto, RAM, Alansohn, Enirac Sum, Mac Davis, Wdfarmer, Shoefly, Gene Nygaard, Ttownfeen, MarcoTolo, Jimgawn, Graham87, Qwertyus, DePiep, Joffan, FlaBot, Nivix, Huntersquid, Physchim62, Chobot, YurikBot, RobotE, Anonymous editor, Hede2000, Buster79, Anareon, Dhollm, PhilipO, E2mb0t~enwiki, Tony1, Elkman, 2over0, Јованб, Pifvyubjwm, Tim R, SmackBot, Hydrogen Iodide, Bluebot, MalafayaBot, DHN-bot~enwiki, Darth Panda, Ctifumdope, Aldaron, Mion, Lambiam, Jaganath, Mbeychok, Bjankuloski06en~enwiki, Ryulong, IvanLanin, Merrowski, Gosolowe, Stifynsemons, Vaughan Pratt, CmdrObot, Robert Rossi, Howardsr, Shirulashem, Arbitrary username, Saintrain, Runch, Nonagonal Spider, Dtgriscom, Greg L, Escarbot, Hekerui, Web-Crawling Stickler, Brianvon, Jackson Peebles, MartinBot, J.delanoy, Hodja Nasreddin, Silas S. Brown, I310342~enwiki, WilfriedC, DorganBot, VolkovBot, Spurius Furius Fusus, Postie77, Andy Dingley, Alexanderhowell, Brianga, AlleborgoBot, EmxBot, Neparis, Kbrose, SieBot, The way, the truth, and the light, Prillen, Flyer22 Reborn, Radzewicz, ClueBot, Polyamorph, Alexbot, Dezaxa, Friend-Of-No-ONe, Addbot, Some jerk on the Internet, Lingosalad, Ronhjones, Download, Rterwer, Wakeham, Numbo3-bot, Lightbot, Luckas-bot, Yobot, AnomieBOT, Daniele Pugliesi, Wikifor, Xqbot, Esmu Igors, Jack B108, Plasticspork, TobeBot, Blind cyclist, Baue0242, 777isHARDCORE, J. Garai, EmausBot, KHamsun, Bro2kno, Fæ, 1howardsr1, ClueBot NG, Xinleiucd, Sprinkler21, Henneyj, Sandbh, Nelg, Aisteco, Dexbot, Czech is Cyrillized, Philadelphus, Cnrowley, I Ðb, AshtonHM, Kent5915, Latosh Boris and Anonymous: 141

• **Latent heat** *Source:* https://en.wikipedia.org/wiki/Latent_heat?oldid=698351023 *Contributors:* Jdpipe, Patrick, William M. Connolley, Emperorbma, Fcrozat, Robbot, Hadal, Wereon, Jleedev, Unfree, Alan Liefting, Giftlite, Zigger, Hokanomono, Tagishsimon, Utcursch, Pgan002, Yardcock, Karol Langner, Eranb, Dylan Shell, Mormegil, NathanHurst, Discospinster, Vsmith, Shadow demon, Rbj, Enirac Sum, Riana, AndreasPraefcke, Gene Nygaard, Alai, Aza~enwiki, Isnow, Hard Raspy Sci, DocRuby, FlaBot, DVdm, YurikBot, Alethiareg, NawlinWiki, Ojcit, Dhollm, Diotti, Lockesdonkey, Pegship, U.S.Vevek, E Wing, Tomj, SmackBot, InverseHypercube, Unyoyega, Gilliam, Bluebot, Pieter Kuiper, Silly rabbit, Baa, Addshore, Salthebad, SundarBot, Mwtoews, Sadi Carnot, Lambiam, Goodnightmush, Nonsuch, Pflatau, Werdan7, RMHED, Runningonbrains, Fl, Quibik, Soumya.92, Thijs!bot, Epbr123, Wikid77, Pjvpjv, Mentifisto, Lm13700~enwiki, JAnDbot, Transce080, VoABot II, Loonymonkey, Ac44ck, 0612, MartinBot, R'n'B, Vvitor, I310342~enwiki, Steve.kimberley, VolkovBot, Fbifriday, Rei-bot, Jennjenn99, Yk Yk Yk, Manikato, Falcon8765, Enviroboy, Kbrose, SieBot, AquaDTRS, Ødipus sic, Tombomp, Kaboooz, Florentyna, Pinkadelica, Denisarona, ClueBot, Binksternet, The Thing That Should Not Be, Helenabella, Niceguyedc, Djr32, Gnunesjr, Antinoah, Nathan Johnson, Addbot, Willking1979, AndrewHarvey4, Elmondo21st, MagnusA.Bot, Steve1515, D.c.camero, Tide rolls, Lightbot, MuZemike, Luckas-bot, AnomieBOT, Rubinbot, Daniele Pugliesi, Jim1138, Piano non troppo, Unara, Citation bot, GB fan, Xqbot, Zad68, Pablovaldes, Pontificalibus, RibotBOT, Mathonius, A. di M., Chjoaygame, S300m, Jets84, I dream of horses, Minored, Jaochainoi, MastiBot, Mikewarbz, Weemallon, Tmondol, Stuey85, Suffusion of Yellow, Khpatil, NerdyScienceDude, Tommy2010, Akhil 0950, Isaackwok, Yerocus, L Kensington, ChuispastonBot, RockMagnetist, ClueBot NG, MelbourneStar, Movses-bot, Esnascosta, Lolm8, Summir, Shikhars24, Mariraja2007, CarrieVS, Lugia2453, Me, Myself, and I are Here, Faizan, Mistystix, Tentinator, Jim Sizemore, Jianhui67, Inyafase, MrWooHoo, Robevans123, Trackteur, Biblioworm, Agoel5920, Shruthi20314, FourViolas, Sanath 100 and Anonymous: 225

• **Latent internal energy** *Source:* https://en.wikipedia.org/wiki/Latent_internal_energy?oldid=696141249 *Contributors:* Bearcat, Rich Farmbrough, Xezbeth, Dhollm, JohnCD, Manuelkuhs, Cydebot, Jarry1250, Addbot, Yobot, Daniele Pugliesi, Jim1138, JimVC3, Erik9bot, Hhhippo, CarrieVS, The Quixotic Potato and Anonymous: 4

• **Trouton's ratio** *Source:* https://en.wikipedia.org/wiki/Trouton'{}s_ratio?oldid=689187044 *Contributors:* Charles Matthews, Rich Farmbrough, Linas, X42bn6, Dhollm, DogFog, Jj137, Rod57, JL-Bot, Wolfch, Addbot, Luckas-bot, Daniele Pugliesi, زادگان قلی, Erik9bot and Anonymous: 2

• **Volatility (chemistry)** *Source:* https://en.wikipedia.org/wiki/Volatility_(chemistry)?oldid=697197041 *Contributors:* Ellywa, Giftlite, Discospinster, Vsmith, Alansohn, Walkerma, Commander Keane, JRHorse, KKramer~enwiki, Carrionluggage, Erik4, YurikBot, Buster79, Trovatore, Dhollm, Poppy, Infinity0, Cmglee, Itub, FocalPoint, Victor M. Vicente Selvas, GregRM, Nbarth, Colonies Chris, Mbeychok, GetsEclectic, LouisBB, Rifleman 82, Christian75, Epbr123, Headbomb, Marek69, Pfranson, Txomin, WilfriedC, Comrade Tux, Kbrose, PlanetStar, Faradayplank, Simkiss3, JL-Bot, Leighwaldon, ClueBot, Goodvac, Addbot, Tide rolls, Yobot, Aboalbiss, Daniele Pugliesi, Citation bot, قلی زادگان, Jonesey95, Double sharp, Sumone10154, Vekov, Katherine, ה אריה., Wikipelli, RaymondSutanto, Δ, ClueBot NG, MerlIwBot, BattyBot, Prouder Mary, YiFeiBot, CogitoErgoSum14, Monkbot, Crossark and Anonymous: 48

• **Binodal** *Source:* https://en.wikipedia.org/wiki/Binodal?oldid=690036343 *Contributors:* 99of9, Dhollm, Pegship, CommonsDelinker, Martinor, Locke9k, WereSpielChequers, Addbot, Luckas-bot, Yobot, ArthurBot, Staszek Lem, Lorem Ip, BG19bot, Going Turbo and Anonymous: 2

• **Compressed fluid** *Source:* https://en.wikipedia.org/wiki/Compressed_fluid?oldid=660332261 *Contributors:* Berek, Firebird, Bearcat, Nsaa, Walkerma, Shoefly, Rjwilmsi, Titoxd, Tedder, Siddhant, Dhollm, Pdcook, Vipinhari, Flyer22 Reborn, ClueBot, Ariadacapo, SchreiberBike, ZooFari, MystBot, Xqbot, Edderso, ClueBot NG, Helpful Pixie Bot, CaSJer, Imran Jahid and Anonymous: 14

• **Cooling curve** *Source:* https://en.wikipedia.org/wiki/Cooling_curve?oldid=671469941 *Contributors:* Jdpipe, Fredrik, MarkSweep, Pedant, Josh Parris, Skz~enwiki, Nopol10, Mysid, Pegship, SmackBot, Davepape, Jrockley, Shalom Yechiel, Robofish, Chuy1530, Wizard191, Thijs!bot, AntiVandalBot, .anacondabot, Barneca, TenTonParasol, Anticipation of a New Lover's Arrival, The, Addbot, Luckas-bot, Wikipelli, ClueBot NG, Sahirbutt and Anonymous: 12

• **Equation of state** *Source:* https://en.wikipedia.org/wiki/Equation_of_state?oldid=694088257 *Contributors:* Tobias Hoevekamp, CYD, Bryan Derksen, Ap, Stokerm, Peterlin~enwiki, Jdpipe, Patrick, Michael Hardy, MartinHarper, Andres, Charles Matthews, Stismail , Krithin, ThomasStrohmann~enwiki,Echoray, Topbanana, Bloodshedder, PuzzletChung, Robbot, COGDEN, Robinh, Tobias Bergemann, Connelly, Giftlite, Graeme Bartlett, Drat-man, Suspekt~enwiki, Mboverload, H Padleckas, Guanabot, Pmetzger, Andreww, NetBot, Duk, Scentoni, Helix84, Alansohn, Spectre630,PAR, SMesser, Malo, Shoefly, Gene Nygaard, Oleg Alexandrov, Thryduulf, Benbest, Rend~enwiki, Vanished user 05, Zzyzx11, Joke137,Graham87, Jan van Male, Sumanch, Eubot, Jfiling, Bgwhite, Roboto de Ajvol, YurikBot, NTBot~enwiki, Hillman, Gaius Cornelius, David-Conrad, Dhollm, Tony1, Necronomicon~enwiki, Kkmurray, SmackBot, Alan Pascoe, Vald, KocjoBot~enwiki, Nscheff ey, Sundaryourfriend,Kmarinas86, Hugo-cs, Kurykh, Complexica, Colonies Chris, Ascentury, Berland, E4mmacro, G 716, Sadi Carnot, Lambiam, L0rents, Mb-eychok, 16@r, Knights who say ni, Mets501, DGtal, IvanLanin, Wquester , Cydebot, Freak in the bunnysuit, Kablammo, Headbomb, TomBarlow, AntiVandalBot, Fayenatic london, JAnDbot, MER-C, Magioladitis, Baccyak4H, Mythealias, STBot, Tgeairn, Decaheximal, Salih,

P.wormer, Gogobera, A4bot, Brking, Venny85, Neparis, Kbrose, Natox, Hiddenfromview, RedHotIceCube, ClueBot, BenXM8, Bbanerje, BOTarate, NoamTene, Crowsnest, DumZiBoT, RP459, MystBot, Mortense, DOI bot, Lightbot, Zorrobot, Legobot, Yobot, Groucho NL, Maltelauridsbrigge, Synchronism, AnomieBOT, Daniele Pugliesi, Materialscientist, Xenomancer, Citation bot, Maniadis, Xqbot, J JMesserly, Omnipaedista, Doulos Christos, Stieltjes, FrescoBot, Citation bot 1, Geraldo62, Rruffpaw, RjwilmsiBot, AndyHe829, EmausBot, Wikitanvir-Bot, John of Lancaster, Hssguy, R. J. Mathar, SporkBot, Maschen, ClueBot NG, CharlesCo, BG19bot, F=q(E+v^B), Smarandi, RudolfRed, ChrisGualtieri, EagerToddler39, Subrata0205, Tiffiw~enwiki, Petronerd, Thermodynama, Loraof, Varsik500, Arghya Chakraborty (Mathematician), Mattyiceyo, WBHolzapfel and Anonymous: 154

- **Leidenfrost effect** *Source:* https://en.wikipedia.org/wiki/Leidenfrost_effect?oldid=692630040 *Contributors:* Alex.tan, Arvindn, PierreAbbat, Theresa knott, Samw, Charles Matthews, AnonMoos, Pakaran, RadicalBender, Naddy, Hadal, Xanzzibar, Achurch, Wolfkeeper, Siroxo, Andycjp, LucasVB, Karol Langner, OwenBlacker, Necrothesp, Yorg, Cyclopia, Diamonddavej, R. S. Shaw, Vystrix Nexoth, Hob Gadling, Tcp-ip, Zelda~enwiki, Hooperbloob, Klhuillier, Frodet, Pauli133, MGTom, Saperaud~enwiki, Bensin, The wub, YurikBot, Huw Powell, Chris Capoccia, Chuck Carroll, Hellbus, Malcolma, Dhollm, Varano, Tomj, SmackBot, Jclerman, McGeddon, Praxio, Buck Mulligan, Bluebot, Nbarth, Melbournian, Fuhghettaboutit, Ma.rkus.nl, PeterJeremy, Twigge, User99~enwiki, Eltouristo, Arbustoo, Collect, DAMurphy, Kurtan~enwiki, Morganfitzp, JohnCD, Tingrin87, Escarbot, MoogleDan, Lfstevens, JAnDbot, Magioladitis, Swpb, Flaming Ferrari, BlackPocket, CommonsDelinker, 5theye, Num1dgen, Inwind, Amikake3, Venny85, Andy Dingley, VanBuren, Jehorn, Cryonic07, SieBot, Paradoctor, Da Joe, Cwkmail, AFA pony, HighInBC, Escape Orbit, ClueBot, Dan.hook, Scottstensland, Alexbot, Myceteae, Pichpich, MystBot, Addbot, Masur, Ijmitchell, OlEnglish, Pietrow, Zorrobot, Yobot, Julia W, Amirobot, EnTerr, AnomieBOT, Captain Quirk, Citation bot, Pauldauenhauer, Syberiyxx, Almuhammedi, Tom.Reding, Skyerise, Exodroner, Ffiti, Jfsalazars, EmausBot, Marco Guzman, Jr, ClueBot NG, Rockboy3970, Jdperkins, Bibcode Bot, BG19bot, Bonnie13J, Yowanvista, IluvatarBot, CitationCleanerBot, BattyBot, Suradnik50, Denysbondar, RaphaelR-Barbosa, Clerish, Scoobydunk, Opencooper, DSCrowned, Lateral byte, Treasuredwealth, Knife-in-the-drawer, Abigaildespotovic, Orielno and Anonymous: 71

- **Macroscopic quantum phenomena** *Source:* https://en.wikipedia.org/wiki/Macroscopic_quantum_phenomena?oldid=696073643 *Contributors:* Alan Liefting, Giftlite, DragonflySixtyseven, Gareth Jones, Arthur Rubin, InverseHypercube, Colonies Chris, Maurice Carbonaro, Dragnmn, Benboy00, Yobot, AnomieBOT, Materialscientist, Citation bot, I dream of horses, ArthurPSmith, Physics therapist, Jesse V., Dick Chu, EmausBot, Dewritech, Tls60, Bibcode Bot, Petrarchan47, DarafshBot, 786b6364, Adwaele, Dexbot, Agonbroke, Jharrism2, Yen-Tzu and Anonymous: 14

- **Mpemba effect** *Source:* https://en.wikipedia.org/wiki/Mpemba_effect?oldid=698340800 *Contributors:* AxelBoldt, The Anome, Shii, Maury Markowitz, Heron, Hephaestos, Michael Hardy, Karada, Tregoweth, Ahoerstemeier, William M. Connolley, Dcoetzee, Dysprosia, Furrykef, Toreau, Babbage, Henrygb, JackofOz, Robinh, Xanzzibar, Cutler, Toytoy, Noirum, DragonflySixtyseven, Icairns, Asbestos, Iantresman, O'Dea, Discospinster, Rich Farmbrough, Suppafly, FT2, Cacycle, Vsmith, Rspeer, Pmcm, Causa sui, Viriditas, Cmdrjameson, Ethomsen, I9Q79oL78KiL0QTFHgyc, Zetawoof, Nsaa, Jérôme, Nroets, Arthena, Curious1i, Sciurinæ, Vuo, YixilTesiphon, Undecided, Isnow, Hard Raspy Sci, Rjwilmsi, Jsled, Rklisowski, Syndicate, Ewlyahoocom, Riki, Mathrick, Kri, King of Hearts, YurikBot, Wester, Rxnd, Hede2000, Rsrikanth05, Grafen, Dhollm, Mrwriter, 2over0, Gulliveig, Nolanus, Digfarenough, Eptin, SmackBot, Amatulic, Thumperward, AtmanDave, Nbarth, Quadparty, BullRangifer, Hgilbert, Vina-iwbot~enwiki, Sophia, Gobonobo, VincentH, Andrés D., Optakeover, Kvng, KJS77, Wentu, JForget, Van helsing, Jeanb., Kjknohw, Cydebot, Doug Weller, Christian75, Malleus Fatuorum, Thijs!bot, Martin Hogbin, Headbomb, Dmit-Trix, Shot info, Dawnseeker2000, Escarbot, FireHorse, Bongwarrior, VoABot II, Xb2u7Zjzc32, Gabriel Kielland, DerHexer, Greenguy1090, Momojeng, PMG, Tgeairn, Emitozzi, Ogno, Dmitri Yuriev, Trilobitealive, BernardZ, UnicornTapestry, Shiggity, Hammersoft, Wiae, Staka, RaseaC, Richard A Muller, Blood sliver, Ignoscient, SieBot, Gorpik, Ivan Štambuk, Paradoctor, Flyer22 Reborn, Steven Crossin, Drjamesaustin, Svick, Superbeecat, Dolphin51, Denisarona, ClueBot, Rumping, Sentewolf, The Thing That Should Not Be, L'omo del batocio, Niceguyedc, DaveEngineer, Auntof6, Crywalt, GrahamDo, XLinkBot, Laurenxxx189, Josephfrazier, Martin Chaplin, Master mack, Faerengol, The Tutor, Addbot, Some jerk on the Internet, DOI bot, Blethering Scot, Sheepech, SamatBot, Barak Sh, Ehrenkater, Luckas-bot, Yobot, Drgao, AnomieBOT, Götz, Rockypedia, Jim1138, Kingpin13, Jbyersjcox, Materialscientist, The High Fin Sperm Whale, Citation bot, Arthur-Bot, LilHelpa, The sock that should not be, Omnipaedista, Raulshc, Creation7689, FrescoBot, IO Device, D'ohBot, Berny68, Citation bot 1, Wdcf, Jonesey95, RedBot, Trappist the monk, Eng.ahmed.m.ali, Vrenator, Jfmantis, Syncategoremata, Klbrain, Tommy2010, Blin00, Wikipelli, Dcirovic, Mz7, Josve05a, Claritas, Druzhnik, Libb Thims, Mediapioneer, L Kensington, Donner60, ClueBot NG, Livefreeordie22, Anagogist, Magpieram, Widr, Antiqueight, Helpful Pixie Bot, Bibcode Bot, BG19bot, AndrePooh, Zompenguin, Lukys~enwiki, Mattwc04, Bsmith2414, Aisteco, Ckwood, ChrisGualtieri, Sbtxag13, Epicgenius, Kno333, François Robere, Xuigh, Haminoon, Loniux, Paul2520, Posh-Frosh, Mukulkumar12345, A.gholami, Monkbot, Nateowami, Mrlybra1987, SamWilson989, Caden Mitchell, Ecqsun, United101united, OkieRick and Anonymous: 265

- **Order and disorder (physics)** *Source:* https://en.wikipedia.org/wiki/Order_and_disorder_(physics)?oldid=638818968 *Contributors:* Charles Matthews, Phys, Smalljim, Pearle, Anthony Appleyard, Coma28, BradBeattie, Nikkimaria, SmackBot, Reedy, QFT, Cybercobra, Smartcat, Dilane, Maliz, Superkuh, Likebox, JL-Bot, Mild Bill Hiccup, SchreiberBike, AnonyScientist, Hess88, Addbot, Mjamja, Legobot, KamikazeBot, Ufim, Xqbot, Erik9bot, Freddy78, Marie Poise, RjwilmsiBot, EmausBot, ClueBot NG, Shaun, Mglbby5, Cosmicraga and Anonymous: 14

- **Spinodal** *Source:* https://en.wikipedia.org/wiki/Spinodal?oldid=681535664 *Contributors:* Askewchan, Dhollm, Martinor, Pdcook, Locke9k, Sean.hoyland, Addbot, The Sage of Stamford, Lorem Ip, BG19bot, Jonathancarroll.hull and Anonymous: 3

- **Superconductivity** *Source:* https://en.wikipedia.org/wiki/Superconductivity?oldid=698397737 *Contributors:* AxelBoldt, Lee Daniel Crocker, CYD, Bryan Derksen, AstroNomer~enwiki, Andre Engels, DavidLevinson, Quintanilla, Jqt, Azhyd, Waveguy, David spector, Heron, Olivier, Edward, Michael Hardy, Fred Bauder, DopefishJustin, Dominus, Karada, Tiles, Egil, Ahoerstemeier, Stevenj, Theresa knott, Snoyes, Julesd, Glenn, Cimon Avaro, GCarty, Cryoboy, Mxn, Tantalate, Reddi, Stone, Joerg Reiher~enwiki, Hao2lian, DJ Clayworth, E23~enwiki, Furrykef, Taxman, LMB, Fibonacci, Omegatron, Traroth, Tophanana, Pstudier, Pakaran, Phil Boswell, Donarreiskoffer, Robbot, Stephan Schulz, Rorro, Bkell, Hadal, UtherSRG, Robinh, Diberri, Cyberpunks~enwiki, Connelly, Giftlite, DocWatson42, MarkPNeyer, Harp, Tom harrison, Ferkelparade, Fastfission, Xerxes314, Leonard G., Foobar, Bobblewik, Wmahan, Irarum, Geni, Quadell, Spiralhighway, Icairns, Peter bertok, Gerrit, Deglr6328, Deeceevoice, Moxfyre, Reflex Reaction, Zowie, CALR, Discospinster, FT2, Rama, Vsmith, Pavel Vozenilek, Paul August, Andrejj, Kaisershatner, CanisRufus, Kwamikagami, PhilHibbs, Haxwell, Simonbp, Femto, Dalf, Bobo192, BrokenSegue, Enric Naval, Slicky, Kjkolb, Nk, Merope, PaulHanson, GiantSloth, Lightdarkness, Sligocki, Pion, Hu, Velella, Wtshymanski, Evil Monkey, RJFJR, Cmapm,

Dfalkner, Gene Nygaard, Aeronautics, RHaworth, Dandv, StradivariusTV, Oliphaunt, Jeff3000, Jwanders, Alfakim, Firien, Triddle, Someone42, GregorB, Eras-mus, CharlesC, SeventyThree, Christopher Thomas, Graham87, Magister Mathematicae, Jan van Male, Josh Parris, Sjö, Sjakkalle, Rjwilmsi, Seidenstud, Fish and karate, FlaBot, PhilipSargent, Jeepo~enwiki, Gurch, Leslie Mateus, Fosnez, Goudzovski, Skierpage, Chobot, DVdm, Ahpook, Takaaki, Roboto de Ajvol, The Rambling Man, Wavelength, Mollsmolyneux, Bhny, JabberWok, Netscott, Hydrargyrum, CambridgeBayWeather, Salsb, GeeJo, Harksaw, Długosz, RyanLivingston, Ino5hiro, Mkouklis, Nineteenthly, Mccready, Dhollm, Scottfisher, Quarky2001, DeadEyeArrow, Oliverdl, Tonym88, Codell, Searchme, Light current, 2over0, DaveOinSF, Theda, Closedmouth, Bamse, Filou~enwiki, Petri Krohn, JoanneB, Alias Flood, Wylie440, Chaiken, SkerHawx, Kgf0, Children of the dragon, SmackBot, Melchoir, Gilliam, Oscarthecat, Chaojoker, Kmarinas86, Chris the speller, RevenDS, NCurse, Thumperward, Papa November, Complexica, AtmanDave, Kostmo, Dual Freq, Trekphiler, KaiserbBot, TheKMan, LouScheffer, Elendil's Heir, Toomontrangle, Pwjb, Smokefoot, Eynar, DMacks, Paulish, Simon Arnold, Lester, Nbishop, Breadbox, Kuru, John, JorisvS, Smartyllama, Manjish, IronGargoyle, Spiel496, Citicat, Majormcmuffin, Kvng, Astrobradley, JarahE, KJS77, Brienanni, Japhet, Hmtamza, Tawkerbot2, Chetvorno, CmdrObot, Van helsing, MorkaisChosen, CBM, WMSwiki, Tim1988, Lokal Profil, Phatom87, Britannic~enwiki, Cydebot, Kam42705, Neil Froschauer, Chasingsol, Myscrnnm, Lee, IComputerSaysNo, Arwen4014, Editor at Large, TrevorRC, Matwilko, Raschd, Epbr123, Kubanczyk, Dasacus, Headbomb, Dgies, Cyclonenim, Courtjester555, Mojohaza1, Casomerville, Yellowdesk, JAnDbot, Quentar~enwiki, Smartcat, Bongwarrior, VoABot II, Ginga2, SineWave, Jjasi, Web-Crawling Stickler, Dirac66, Coolkoon, Limtohhan, Joshua Davis, Schmloof, Xantolus, CommonsDelinker, Pharaoh of the Wizards, Jtw11, Dmrmatt19, Hans Dunkelberg, Uncle Dick, Maurice Carbonaro, Nigholith, MrBell, Eliz81, Bakkouz, Rod57, Bot-Schafter, TomyDuby, Anatoly larkin, Wimox, Equazcion, Tevonic, Useight, Qaz123qaz, Bertiethecat, Idioma-bot, JeffreyRMiles, VolkovBot, TXiKiBoT, Neha simon, Calwiki, Hqb, Liquidcentre, JosephJohnCox, OlavN, Sodapopinski, Robert1947, Burntsauce, Elecwikiman, Fischer.sebastian, AlleborgoBot, Shanmugampl, Runewiki777, Steven Weston, SieBot, Yintan, Vanished User 8a9b4725f8376, FSUlawalumni, Keilana, Hzh, Henry Delforn (old), Onopearls, Anchor Link Bot, Hamiltondaniel, Geoff Plourde, Elliott-rhodes, TubularWorld, Tegrenath, LarRan, ClueBot, Trojancowboy, Fuzzylunkinz, Ctiefel, Techdawg667, VsBot, YBCO, Niceguyedc, Rotational, Cousins.inc, CohesionBot, Jeck1335, Doctorpsi, PixelBot, Bob man801, Lartoven, Brews ohare, Neucleon, Natty sci~enwiki, Doprendek, SchreiberBike, Aitias, Subash.chandran007, SoxBot III, HumphreyW, LSTech, Tarlneustaedter, Wertuose, BodhisattvaBot, Rror, Ngebbett, Ost316, WikHead, Noctibus, ElMeBot, Addbot, Forscite, AVand, DOI bot, Melab-1, Travisoto, Flning, Jncraton, CanadianLinuxUser, Leszek Jańczuk, CarsracBot, Dr. Universe, K Eliza Coyne, Gwcdt, Lightbot, SPat, Luckas-bot, Yobot, Fraggle81, THEN WHO WAS PHONE?, CinchBug, Csmallw, MassimoAr, AnomieBOT, Cryogenics, Guff2much, Materialscientist, Citation bot, Xqbot, Eep not for fat people, Waleswatcher, NinjaDreams, Janolaf30, Dave3457, GliderMaven, FrescoBot, WikiMcGowan, Tobby72, AlanDewey, Citation bot 1, ASchwarz, Pinethicket, HRoestBot, Schrodingers rabbit, 10metreh, Tom.Reding, Gruntler, Richardc03, Mikespedia, Heller2007, Felix0411, Anoop ranjan, Aleitner, Ahsbenton, Agnel P.B., Catcamus, Akoufos, Jiyojolly, Dick Chu, Noommos, Haj33, EmausBot, John of Reading, WikitanvirBot, DonyG, JasonSaulG, Mathew10111, Pascalf, Hhhippo, ManosHacker, Medeis, A930913, Tls60, Sailsbystars, Nothingbutdreamer, ChuispastonBot, AndyTheGrump, DASHBotAV, WikiBaller, ClueBot NG, Gilderien, Hightc, Widr, Names are hard to think of, Helpful Pixie Bot, Sina.zapf, Mightyname, Nightenbelle, Jubobroff, Bibcode Bot, BG19bot, Virtualerian, Island Monkey, Ymblanter, Andol, WikiHacker187, Mark Arsten, 52 6f 62, Pong711, BattyBot, Bv.vasiliev, Chim02, MahdiBot, Jimw338, Embrittled, Adwaele, Protectionwi, Dexbot, Anandaraja, Oliver brookes, Fittold27, TwoTwoHello, Andyhowlett, Reatlas, Ruby Murray, François Robere, Rabbitflyer, Asik Ram, Monkbot, Jrafner, Laurencejwolf, Scipsycho, OzRamos, KasparBot, Superspin, Shao xc and Anonymous: 526

• **Superheated water** *Source:* https://en.wikipedia.org/wiki/Superheated_water?oldid=689757405 *Contributors:* Stone, Orpheus, PFHLai, Rich Farmbrough, Bender235, Smalljim, Woohookitty, Bgwhite, Dhollm, BorgQueen, Slashme, Edgar181, Chris the speller, Thumperward, Nahum Reduta, Richard001, Iridescent, TheTito, Christian75, Helvetica, Lfstevens, Nono64, Tmorgan7, TXiKiBoT, Lamro, Gbeebani, Lightmouse, Shniken1, Sting au, Mr squelch, Stainless316, Doprendek, Thewellman, Plasmic Physics, Addbot, Mortense, DOI bot, $4projects, Götz, FrescoBot, Citation bot 1, Full-date unlinking bot, Trappist the monk, H3llBot, Diamondust0, ClueBot NG, Delusion23, Aniketmh035569, Helpful Pixie Bot, MusikAnimal, BattyBot, Editfromwithout, Spyglasses, Monkbot, CLCStudent and Anonymous: 21

• **Superheating** *Source:* https://en.wikipedia.org/wiki/Superheating?oldid=692984871 *Contributors:* Edward, Notheruser, Glenn, Schneelocke, Mulad, Dcoetzee, EACH, Tempshill, Morven, Hankwang, Plasmaroo, Buster2058, ShaunMacPherson, Amp, Wmahan, DragonflySixtyseven, MementoVivere, Deglr6328, Vsmith, Chewie, J-Star, Jhd, Thorpe, Bkuschel, Jeff3000, JohnJohn, Nneonneo, YurikBot, Mushin, Chuck Carroll, Hellbus, Dhollm, SmackBot, Blue520, Jmurphy914, Jushi, Thumperward, Invenio, Can't sleep, clown will eat me, Nipahc, Richard001, Stefano85, The Ungovernable Force, Vindictive Warrior, Flip619, Lim Wei Quan, Hotmop, Spiel496, SquidThing, Tawkerbot2, Heqs, CmdrObot, Insanephantom, A876, Headbomb, Mysterial, DarthShrine, BeefRendang, Swpb, Beagel, F3et, Silas S. Brown, TWCarlson, VolkovBot, Kyle the bot, TXiKiBoT, Ciscokid21, Marquetry28, Lightmouse, Isatube, Chokoboii, Snigbrook, Johnyapchapco, VQuakr, Stainless316, Gmxten, DumZiBoT, XLinkBot, Kbdankbot, Addbot, Hda3ku, Bader A. Shehab, Luckas-bot, AnomieBOT, Rubinbot, Xqbot, CXCV, Srich32977, Muion, Mfwitten, HRoestBot, Tom.Reding, ZéroBot, Mac LAK, Porter157, L Kensington, Jeff Song, Ciro.santilli, Helpful Pixie Bot, Northamerica1000, Alexwho314, Blackbombchu, Hamedhashemi60, Himanshu4777 and Anonymous: 72

• **Thermo-dielectric effect** *Source:* https://en.wikipedia.org/wiki/Thermo-dielectric_effect?oldid=609127199 *Contributors:* Stone, Topbanana, Dhollm, Hmains, OrenBochman, Magioladitis, Addbot, Citation bot, Marshallsumter, Citation bot 1, Martintomas, Helpful Pixie Bot and Bibcode Bot

21.3.2 Images

21.3.3 Content license

www.ingramcontent.com/pod-product-compliance
Lightning Source LLC
Chambersburg PA
CBHW080703190526
45169CB00006B/2224